Anarchy and Geography

This book provides a historical account of anarchist geographies in the UK and the implications for current practice. It looks at the works of Frenchman Élisée Reclus (1830–1905) and Russian Pyotr Kropotkin (1842–1921) which were cultivated during their exile in Britain and Ireland.

Anarchist geographies have recently gained considerable interest across scholarly disciplines. Many aspects of the international anarchist tradition remain little-known and English-speaking scholarship remains mostly impenetrable to authors. Inspired by approaches in historiography and mobilities, this book links print culture and Reclus and Kropotkin's spheres in Britain and Ireland. The author draws on primary sources, biographical links and political circles to establish the early networks of anarchist geographies. Their social, cultural and geographical context played a decisive role in the formation and dissemination of anarchist ideas on geographies of social inequalities, anti-colonialism, anti-racism, feminism, civil liberties, animal rights and 'humane' or humanistic approaches to socialism.

This book will be relevant to anarchist geographers and is recommended supplementary reading for individuals studying historical geography, history, geopolitics and anti-colonialism.

Federico Ferretti is a Lecturer in Human Geography at University College Dublin, Ireland. He discussed a PhD dissertation on Élisée Reclus's *New Universal Geography* at the Universities of Bologna and Paris 1 Panthéon-Sorbonne. He has taught in Italy, Switzerland, France and Brazil. His main research interests lie in alternative geographical traditions and the international and multilingual circulation of geographical knowledge, especially from Latin America and continental Europe.

Routledge Research in Historical Geography

Series editors:
Simon Naylor, *University of Glasgow, UK* and **Laura Cameron**, *Queen's University, Canada*

This series offers a forum for original and innovative research, exploring a wide range of topics encompassed by the sub-discipline of historical geography and cognate fields in the humanities and social sciences. Titles within the series adopt a global geographical scope and historical studies of geographical issues that are grounded in detailed inquiries of primary source materials. The series also supports historiographical and theoretical overviews, and edited collections of essays on historical-geographical themes. This series is aimed at upper-level undergraduates, research students and academics.

Cultural Histories, Memories and Extreme Weather
A Historical Geography Perspective
Edited by Georgina H. Endfield and Lucy Veale

Commemorative Spaces of the First World War
Historical Geographies at the Centenary
Edited by James Wallis and David C. Harvey

Architectures of Hurry—Mobilities, Cities and Modernity
Edited by Phillip Gordon Mackintosh, Richard Dennis, and Deryck W. Holdsworth

Anarchy and Geography
Reclus and Kropotkin in the UK
Federico Ferretti

Twentieth Century Land Settlement Schemes
Edited by Roy Jones and Alexandre M. A. Diniz

For more information about this series, please visit: www.routledge.com/Routledge-Research-in-Historical-Geography/book-series/RRHGS

Anarchy and Geography
Reclus and Kropotkin in the UK

Federico Ferretti

Routledge
Taylor & Francis Group

LONDON AND NEW YORK

First published 2019
by Routledge
2 Park Square, Milton Park, Abingdon, Oxon OX14 4RN

and by Routledge
52 Vanderbilt Avenue, New York, NY 10017

First issued in paperback 2020

Routledge is an imprint of the Taylor & Francis Group, an informa business

British Library Cataloguing in Publication Data
A catalogue record for this book is available from the British Library

Library of Congress Cataloging in Publication Data
A catalog record has been requested for this book

ISBN 13: 978-0-367-58776-5 (pbk)
ISBN 13: 978-1-138-48812-0 (hbk)

Typeset in Times New Roman
by Taylor & Francis Books

This book is dedicated to my father Angelo Ferretti (Reggio Emilia, 1930–2018), witness of the anti-fascist memories of my family and my land, and one of the last representatives of generations of self-educated proletarians who first experienced reading books in the night, after long days of manual work. It was from him that I heard the name of Reclus for the first time.

Contents

Figures

Introduction

Alternative geographical traditions

The ambition for this book is that it marks a milestone in the current rediscovery of anarchist geographies by a wide range of scholarly literature in several languages, and especially of what Simon Springer has called *The Anarchist Roots of Geography*. [1] These works demonstrated that anarchy and geography are associated in at least two ways. First, there have been historical correspondences between the two terms, because some of the international 'Founding Fathers' of the anarchist movement were concurrently world-renowned geographers, including Élisée Reclus (1830–1905) and Pyotr Kropotkin (1842–1921). Second, anarchism and (critical) geography can both be considered as specific ways to understand and transform the world giving special emphasis to its spatialities; this is also demonstrated by the variety of fields in which anarchist thinking is still inspiring engaged approaches to geography. [2] All this stands in opposition to the Marxist tradition; unlike Reclus and Kropotkin, Marx and Engels despised geography and neglected spatial thinking as a possible heuristic approach. Despite decades of efforts, albeit generous, to construct Marxist geographies, the recent public debate between Springer and David Harvey shows how Marxism still struggles to adapt its classical conceptual frameworks to geography. [3] That debate provided further support for authors who suspect that the term 'Marxist geography' seems an oxymoron. [4] However, defining anarchist geographies requires methodological caution, since in Reclus and Kropotkin's time, the label 'anarchist geographies' did not yet exist. [5] Therefore, to avoid anachronisms, I refer to these two authors as 'anarchist geographers' and use the definition 'anarchist geographies' only as a reference to concepts that can be associated with contemporary debates that explicitly assume this description. Likewise, I always discuss 'anarchist geographies' in the plural because anarchists never understood geography as entailing the single viewpoint of one party, which was consistent with their traditional claims for the integration of manual and intellectual labour, against the idea of the organic intellectual. Therefore, it is in its plurality that the description 'anarchist geographies' needs to be understood.

The research timeframe here spans from 1852, the year the Reclus brothers arrived in London, to 1917, when Kropotkin left Britain. Reclus, a Frenchman, sought refuge in Great Britain and Ireland from 1852 following the 1851 coup d'état by Louis-Napoléon Bonaparte (later Napoléon III), together with his elder brother Élie Reclus (1827–1904), likewise an anarchist, geographer and anthropologist. They both maintained a constant correspondence with scientists and activists in the lands of their exile all their lives. Kropotkin, a Russian prince exiled for his opposition to the Czar's regime, lived in England at various times from 1876 to 1882, and permanently from 1886 to 1917, becoming a central figure within local political and scientific fields. Nevertheless, with the notable exception of Gerry Kearns' works on Halford Mackinder, geopolitics and the political relevance of geography,[6] there has been little substantive research conducted by geographers on the British and Irish networks of Kropotkin and Reclus. This book addresses for the first time the operation of their English-speaking international professional and political networks as well as the effects of their localisations, cultural transfers and immediate receptions in the 'British Isles'. Conscious of the problems that identifying this geographical area entails, I use the term 'British Isles' rather than other terms such as the 'Anglo-Celtic Islands', because it matches the one used at that time, in English and in other languages (*Iles britanniques*, in Reclus' *Nouvelle Géographie universelle*—hereafter *NGU*). In the book's title, the 'UK' refers to the United Kingdom of Great Britain and Ireland as it existed from 1801 (the Act of Union) to 1922 (Irish independence), a period which more than covers the chronological range of my research. Ireland is therefore included in my discussion, although the 'Celtic Island' interested early anarchist geographers in different ways than England or Scotland, as I subsequently explain.

The ambition for this book is also that it extends and brings into communication several strands of geographical, historical and interdisciplinary scholarship. First, I argue for the relevance of what I call alternative geographical traditions, drawing upon David Livingstone's celebrated work *The Geographical Tradition*.[7] While Livingstone demonstrated the plurality of the possible traditions in geography, I would contend that geographical scholarship still needs to fully take into account the richness and importance of dissident and unorthodox geographies despite some invaluable introductory works on these topics.[8] In a paper recently published by *Progress in Human Geography*, Innes Keighren observed that, after doing great work on what has generally been considered as the despicable side of the geographical tradition (for example, the involvement of geographers with Nazism and Nazi death camps), historical geographers have been increasingly rediscovering the 'admirable' aspects of their history, including that represented by early anarchist geographers.[9] Despite this, the task suggested by David Stoddart in 1986 still needs to be accomplished, that is, to deepen understanding of early critical scholars who strove for 'humanizing the new geography'.[10] In this book, I pay special attention to the concept of humanisation, and to the humane and humanistic aspects of the discipline, suggesting that they are indispensable to fully understanding the relationships between Reclus, Kropotkin and British science and politics.

Substantial work has been recently undertaken on historical geographies of anti-colonialism and decolonisation,[11] as well as on historical geographies of labour and global solidarity.[12] Drawing upon this literature, I strongly contend that engaging with solidarity and sociability networks[13] is likewise necessary to understand the works of early anarchist geographers. Recent research based on primary sources has shown in persuasive ways that theorising by Reclus, Kropotkin and their closest collaborators such as Élie Reclus, Lev Ilič Mečnikov/Léon Metchnikoff (1838–1888), Charles Perron (1837–1909) and Paul Reclus (1858–1941), must be considered as resulting from the efforts of a collective network and cannot be taken as the outcome of one or more eminent individuals.[14] For instance, the anarchist concept of mutual aid, popularised by Kropotkin and inspired by Russian scientific traditions,[15] was in fact the result of a common enterprise that Kropotkin, Reclus and Metchnikoff were involved with in Switzerland in the 1880s while editing the *NGU*.[16] For that reason, I consider Reclus's and Kropotkin's works and networks in the British Isles as closely interconnected; where Reclus can be found, at a certain point you will find Kropotkin, and vice versa. The newly available sources I analyse provide further evidence in support of these assumptions. Therefore, as an original extension of earlier works on Reclus[17] and of intellectual and political biographies of Kropotkin,[18] I assume the primacy of (sociability and solidarity) network effects in understanding their common work. I argue that the networks of early anarchist geographers, and their interaction with wider networks of scholars and activists, form a key element allowing for an adequate assessment of the coherence between anarchy and geography in the writings and political actions of their participants. Systematic analysis of these networks through primary sources shows that political, scholarly and editorial matters were always interconnected in the relations between early anarchist geographers, and in their interactions with the wider scholarly and editorial networks that I analyse in this book.

To undertake this analysis, I draw upon the consolidated methodological approach of relational and transnational histories of anarchism,[19] which has recently informed invaluable works revealing the importance of understanding anarchism through the transnational circulation of its activists and their local sociability networks. These analyses have proved to be especially effective in understanding the importance of London which acted as a global hub for anarchist migrants at that time, recently analysed by Constance Bantman and Pietro di Paola.[20] This transnationalist dimension can be connected with an established strand of literature in historical geography arguing the need for analysing localisations, travel and transfers of knowledge, to understand the construction of concepts.[21] This literature also deals with biographies, considered as an important way to understand the contexts of the production of geographical knowledge by exploring lives, relations and concrete experiences of their producers, strands of research inaugurated in the 1970s with the foundation of journals like *Geographers Biobibliographical*

Studies. Whilst authors like Charles Withers highlighted the 'double importance of geography for biography and of the geography of biography',[22] Elizabeth Baigent noticed that a true 'biographical turn' in historical geography substantiated readings which challenged internalist histories generally presenting the subject of geographical knowledge 'as isolated from wider intellectual, economic and political contexts'.[23] Recently, Innes Keighren confirmed that biography still 'has a vibrant presence and continuing importance in scholarship on the history of geography'.[24] Biographical, relational and space-sensitive approaches can be considered as complementary strategies to read early anarchist geographies in their places. Over time, the British Isles became one of the most important places for the elaboration and dissemination of these critical ideas thanks to the mutual exchange and fertilisation of ideas which resulted from the meeting of local scholarly cultures and socialist traditions, and influences derived from the diverse backgrounds of Reclus, Kropotkin and other migrants.

A major contention of this book relates to Reclus's and Kropotkin's editorial networks, on which I focussed a substantial part of my doctoral and postdoctoral works, involving Reclus' French publishers, and a preliminary study on Kropotkin's British publishers.[25] In recent years, historical geographers have developed an increasing interest in print cultures and in the circulations of books and journals as a means for understanding the construction and dissemination of geographical knowledge through its editorial contexts and materialities.[26] Although recent work by Dean Bond has demonstrated the important role of periodical output in conditioning the production of geographical knowledge, not limited to geographies of the book,[27] I would contend that Reclus's and Kropotkin's cases show a close interconnection of book and periodical output for geographies, revealing how a series of papers became books, and papers became brochures, for tasks of popular dissemination. Extending this scholarly approach, I claim that, when analyses of Reclus's and Kropotkin's British editorial networks and acquaintances are compared, the results confirm the continuously shared nature of their scholarly and political endeavours. Moreover, the correspondence between Reclus, Kropotkin and their British publishers and fellow geographers provide new interpretative keys for understanding their popularity within politically more moderate, and even conservative, scholarly milieus. Although personal trust, common sociability networks and a generalised liberal mentality can explain a part of this story, it is now clear that an important motivation for British scholars welcoming Reclus and Kropotkin was material interest. The anarchist geographers' prestige and commercial success in terms of publications was appealing not only to professional publishers, but also to scientific editors and geographers who had much to gain in collaboration with them. This understanding reinforces my previous claims[28] concerning the relations between early anarchist geographers and the publishing industry as being mutually favourable, in effect, a 'bargain', not only in the commercial sense but also in the sense of what Claudio Minca and Franco

Farinelli have defined as the idea of geography involving implicit political strategy.[29] In a nutshell, Kropotkin and Reclus were happy with circulating their work to a wider public rather than solely to activists, and their scholarly publishers were happy with the sales. While this implied the acceptance of some compromises with the capitalist editorial business world, it can be said that, in terms of Reclus's and Kropotkin's 'bargains', geography served anarchy and anarchy served geography.

While this last claim can be seen as moving beyond earlier Kropotkin studies focussed on a more 'retiring' or 'moderate' political position taken by the anarchist prince around the 1890s, likewise questioned in more recent literature,[30] my final contention concerns the influence of geography on militant anarchism and its specificities. Although it is impossible to make a rigid distinction between exclusively scholarly and exclusively activist editorial engagement, a survey of the anarchist and (humanitarian) socialist press to which Kropotkin and Reclus primarily contributed shows that early anarchism was especially sensitive to what is called today intersectionality, that is, a rejection of single-axis approaches for understanding the multidimensionality of oppression and its possible solutions.[31] My work has shown how women's emancipation and women's activism, as well as early claims of sexuality as involving political matters and early defences of what was called same-sex love, were central concerns of *Freedom*, the journal which served as the primary vehicle for Reclus's and Kropotkin's views in Britain. Civil rights and humanitarian issues were likewise addressed in British anarchist milieus, and Reclus's works advocating environmental protection, vegetarianism, animal rights and interspecific solidarity were among the first political writings of the French anarchist geographer to be translated into English. I especially highlight the originality of the contribution that early anarchists provided on issues in respect of anti-colonialism, anti-racism and challenges to Euro-centrism. Although recent literature has shown that, thanks to its transnational nature, anarchism was the first socialistic movement of European origin whose activists took seriously the problem of dialogue with non-European cultures, including indigenous cultures, especially in Africa, Eastern Asia and Latin America,[32] a survey of *Freedom* and other socialist British publications confirms and fosters extended investigation into what I have already considered concerning the early anarchist geographers' anti-racism and anti-colonialism.[33] These new sources reveal Reclus's and Kropotkin's sensitivity to issues concerning decolonisation in Ireland, Eastern Europe and in the extra-European colonies, as well as their radical anticolonial and anti-militaristic opposition to the 1899–1902 South African War and to the jingoist hegemony of those years. Concerned with spreading knowledge and promoting empathy regarding so-called primitive and non-European peoples, geographical works by Kropotkin and the Recluses clearly played a role in fostering a more cosmopolitan consciousness within European working classes.

This book draws on multiple sources, including a systematic and multilingual survey of Reclus's and Kropotkin's publications and of the British scholarly and activist journals to which they contributed, and extensive research on unpublished correspondence held in British archives such as the RGS-IBG Manuscript Collection, the British Library, the Westminster Archives Centre and the National Library of Scotland, as well as in the International Institute of Social History in Amsterdam, in the National Library of Ireland in Dublin, and in the mammoth Kropotkin collection at the State Archives of the Russian Federation in Moscow (Gosudarstvennyi Arkhiv Rossiiskoi Federatsii—GARF), where more than 3,000 folders of Kropotkin's international correspondence are now open to researchers. Among Reclus's and Kropotkin's acquaintances in the British Isles, one finds geographers and cartographers such as John Scott Keltie, Hugh Mill, Henry Bates, Henry Woodward, Augustus Henry Keane, Halford Mackinder, Ernst Ravenstein, Patrick Geddes, John George Bartholomew and Andrew John Herbertson; publishers and editors such as James Knowles, William Skilbeck, John Black, A. Granger Hutt, Hugh Chisholm and Reginald Smith; editorial agents such as George Herbert Perris and Francis Cazenove; scientists such as John Lubbock, Archibald and James Geikie, William Robertson-Smith and Alfred Russel Wallace; social-democrats including Henry Hyndman, Fabians such as George Bernard Shaw, unorthodox and humanitarian socialists such as William Morris, Walter Crane, Joseph Cowen, Havelock Ellis, Henry Salt, James Mavor, the anti-colonialist journalists Guy Aldred and Henry Nevinson and the pioneer of gay rights Edward Carpenter; British and Irish anarchists such as Alfred Marsh, Nannie Florence Dryhurst, Agnes Henry, Charlotte Wilson, Thomas H. Keell and all the members of the *Freedom* group; transnational anarchists who spent many years on British soil as exiles or workers such as Emma Goldman, Max Nettlau, Varlaam Tcherkesoff, Errico Malatesta, Rudolf Rocker, Saul Yanovski and many others; pioneers of feminism such as Josephine Butler and Anne Cobden-Sanderson; and a main protagonist of the Celtic revival, William Butler Yeats, and the writer Oscar Wilde. Analysing the correspondence of Kropotkin and Élisée Reclus, one finds what might be considered as unexpected connections in common, for instance with the English historian of French Protestantism Richard Heath, a life-long friend of both men.

Methodologically, I draw upon recent historical scholarship highlighting the value of archives as a basis for 'studying transnational intellectual history',[34] and on works by historical geographers discussing the relevance of archival' fieldwork for this field of study[35] and in general arguing for the effectiveness of primary sources to understand 'the movement of geographical knowledge between different print forms'.[36] Therefore, I use extensively sources such as published and unpublished correspondence, as I consider these are paramount to understanding sociability networks, social contexts, and the relations of scholars and activists. This is especially the case with Kropotkin and the Recluses, who spent a large part of their working lives corresponding and networking with their collaborators and fellow thinkers, and also considering that a significant amount of their work and correspondence remains unpublished. Engaging with the insights of 'radical history' and reflecting on how 'archives, and other material forms of the past, can inform the present',[37] by doing histories of geography in the public sphere,[38] I use these archives for constructing 'counter-narratives' and 'usable pasts' in a specific way. Considering the discussion, recently addressed by Paul Griffin, between supporters of Jacques Derrida's idea of archives as markers of historical authority, and supporters of Raphael Samuel's concept of the archive as collective history in the making, and drawing on Latour's ideas concerning the actor-network, material agency and centres of calculation,[39] I claim that archives are likewise actors in networks. Although a researcher is the vehicle that allows these material actors to speak, I make no claims to be 'neutral' or allegedly 'objective' to any extent. I speak from my standpoint as an admirer of the early anarchist geographers, and consider it indispensable to declare my positionality within the field of debate in the same sense in which feminist approaches to participatory research consider that a researcher should not claim to be an external observer of a problem, but rather acknowledge their active involvement within it.[40] In this sense, it is necessary to account for my own experiences in trying to bring order to the intrinsic (conceptual prior to material) disorder of the archives.[41] The most original large-scale collection of sources I have analysed in recent years has been the impressive Kropotkin collection at the GARF in Moscow. After my first visit to the GARF in 2009, the idea arose of developing further research work on this immense corpus following completion of my thesis.

In addition to letters in Russian which were beyond the scope of my work, I had available an enormous number of folders containing unpublished correspondence in Western languages such as French and English (more rarely Spanish and Italian), whose exact number is still difficult to calculate. I would assess it as comprising approximately one thousand folders, each one dedicated to a single correspondent and containing from one to several hundred letters. Due to practical challenges arising because of the uncomfortable and time-consuming conditions of work and access to materials at GARF entailed by its distinctive and not readily comprehensive organisational structures (despite the staff's kindness), administrative constraints around organising sufficiently long stays in Moscow involving the arrangement of visas and expenses, and the extraordinarily complicated, slow and expensive process for paying for and ordering photocopies at a distance, I could not view all the folders I would have liked to consult from 2009 onwards. Therefore, I have had to concentrate my efforts on the correspondents whom I considered more relevant or whose folders contained the largest number of letters. Despite these difficulties, the range of Kropotkin's British correspondents constituted a suitable pool of sources for the construction of a documentary corpus for the purposes of this book. It is also worth noting that Kropotkin's correspondence (as with Reclus's) often presents lacunas and asymmetries. For instance, in most of their correspondence one finds that only the letters received, or only the letters sent, have survived. In some cases, there is an abundance of correspondence in one direction and a dearth of correspondence in the other direction. These issues are discussed case by case in the book, involving highlighting openings and limitations of the selected source, and discussing critically the reasons for considering the documents in question.

In the first chapter of the book, I reconstruct some relevant experiences undergone by the brothers Élie and Élisée Reclus in Britain and in Ireland following their initial exile, starting with their arrival in London on 1st January 1852. I particularly focus on the impact which these experiences had on their political and scholarly formation in terms of their direct experiences of industrial pauperism in England and of the Great Famine in Ireland, and by the intellectual articulation resulting from the fertilisation of ideas between their former German education and British scientific debates increasingly heading towards evolutionary approaches on the eve of the 'Darwinian revolution'. I then consider the scholarly and editorial networks they developed in the British Isles in the following decades, which chiefly involved members of the Royal Geographical Society (RGS) for Élisée Reclus, and of the Anthropological Institute of Great Britain and Ireland for Élie Reclus. In both cases, a special reciprocal feeling between British scholars and the Recluses emerged, beyond political differences, encouraged by common engagements in education, evolutionary developments and geographical publishing.

In the second chapter, I address the scholarly editorial networks of Kropotkin, extending and deepening my former work on this subject, especially thanks to new sources identified in Moscow. What these documents particularly reveal is the centrality of editorial business activities in building the scholarly reputation of Kropotkin, which supported his material position in England, opening up for him the doors of the principal publishers and gaining for him the friendship of the most prestigious scholars. While previous research, including my own work, has placed considerable emphasis on British traditions of tolerance, this new research opens the way for reflection on the material reasons why Kropotkin and Reclus enjoyed such open access within British scholarship. This observation may appear as rather banal, given that publishers would be obviously interested in selling copies. However, it is important to bear in mind that this publishing interest also concerned geographers such as Halford Mackinder, whose letters reveal that his displays of friendship towards Reclus and Kropotkin were mainly intended to be of instrumental value in gathering famous names for launching his editorial series within the countries of the world, and which progressively diminished as his reputation began to develop of itself.

In the third chapter, I analyse the geographical output of both Élisée Reclus and Kropotkin concerning the British Isles, especially the relevant chapters of Reclus's major works, the *NGU* and *L'Homme et la Terre*, as well as his *Guide to London*, and Kropotkin's *Fields, Factories and Workshops*. In my view, the importance of this geographical output lies in the authors' attempts to build early social, political and economic geographies of the British Isles. On the one hand, this region of the world had an impact on these early anarchist geographers' understandings of society and space, while on the other hand, their books gave authoritative weight to anarchist ideas on decentralisation, urbanism, empire, poverty and related topics which influenced later scholarship and contributed to spreading knowledge of the British Isles internationally, thanks to the numerous translations and the wide international circulation which these books enjoyed. Indeed, the principle of productive decentralisation, a staple of anarchist thinking, was elaborated by Kropotkin after observing British industries, and Reclus's critiques of European imperialism were greatly indebted to his personal knowledge of British society and scholarship.

In the fourth chapter, I address for the first time the political networks of Reclus and Kropotkin in the British Isles through analysing their correspondence with key activists, and through a systematic survey of the journal *Freedom* from its foundation by Kropotkin and Charlotte Wilson in 1886 to Kropotkin's departure from Britain in 1917 (although Kropotkin progressively broke with the editorial group from 1914). I claim that these sources reveal the close links that anarchist geographers envisaged between activism and geographical scholarship, and the coherence which existed between these fields of activity, in terms of both the topics addressed and the networks involved. Moreover, my survey of *Freedom* demonstrates that the critical commitment of early anarchists was far from being one-dimensional or centred on a single-axis approach. Articles on women's emancipation and works by *Freedom*'s leading female figures such as Charlotte Wilson, Nannie Dryhurst, Agnes Henry, Lilian Wolfe and others, show that the anarchist tradition engaged with feminism and promoted an active role for women much more than what has been previously thought. Articles and correspondence on Ireland show the early anarchists' commitment to the Irish question and original solutions proposed to merge independence and social revolution, especially supported by figures such as Dryhurst. This survey also shows the early radicalness of anarchist anti-colonialism when addressing issues concerning countries beyond Europe. Although the anarchists did not specifically employ the label of anti-colonialism, considering it as part of a wider social liberation, anarchism proved to have been the most radical anti-colonialist and anti-racist European political movement of that time, and geography appears as one of the main inspirations for this critical understanding of European imperialism and colonial crimes.

The fifth chapter develops further this argument by analysing Reclus's and Kropotkin's relations with a heterogeneous range of activists and scholars who can be considered as part of British 'ethical' or 'humanistic' socialism, including former members of the Fellowships of the New Life such as Edward Carpenter and Havelock Ellis, the founder of the Humanitarian League, and editor of *Humanity* and *Humane Review* Henry Salt, as well as ethical socialists in the broadest sense such as William Morris, Walter Crane, Patrick Geddes and James Mavor, and an exceptional figure of multilingual scholar and Christian socialist Richard Heath, who was one of the most important figures among their British friends and correspondents for both Reclus and Kropotkin. I claim that early anarchist geographers, and anarchists in general, were interested in the struggles for civil rights promoted by these colleagues and friends for two principal reasons. First, claims for same-sex love made by Carpenter, for the political value of sex made by Ellis and the Fellowship, against militarism and corporal and capital punishment made by the Humanitarian League, for vegetarianism, animal rights, and protection of the environment made by Salt's *Humane Review*, and for integral education made by Morris, Geddes and others, matched anarchist concerns in so far as they occurred within the grounds of scholarship and social propaganda, and far from political power. Second, all these actors shared common concerns for the liberation of all humankind (and beyond the human species), and not only of a specific social class, which also explains Kropotkin's and Carpenter's commitment to transmit humanistic concerns into scientific fields through the *Humane Science Lectures*. At that time, the British Isles was an extraordinary laboratory for all these endeavours.

Finally, I must acknowledge many friends and colleagues, apologising in advance to those whom I may have omitted from the following list. From the beginning of my research on these source materials, I received financial support from the Centre National de la Recherche Scientifique (CNRS) équipe épistémologie et Histoire de la Géographie (EHGO) in Paris for my first trip to Russia, then from the University of Geneva from 2012 to 2015 in the context of the project Writing the World Differently (FNS div. 1), which allowed me to deepen my knowledge of the international archives of Reclus and Kropotkin. More recently, the University College Dublin (UCD) College of Social Sciences and Law has supported me with two research grants on 'Early anarchist and anti-colonialist geographies' (grant numbers R15517 and R15975) and 'Reclus and Kropotkin in the UK' (R67861). I especially thank Alun Jones for his rereading of my work and for supporting my research work at UCD. For my access to Russian archives I am especially indebted to Sho Konishi, Francesco Benvenuti, Vladimir Alexandrovič Kozlov, Svetlana Slavkova, and Irina Michailovna Galymzianova for my first steps in Moscow, and more recently to Sergey Saytanov, Anton Loshakov, Pascale Siegrist and David Teurtrie. For discussions and insights on the Recluses and their biographies, I acknowledge my great Reclusian friends and colleagues Philippe Pelletier, Patrick Minder, Ronald Creagh, Nicolas Eprendre, Thierry de Bresson and Christophe Brun. For the idea of investigating British publishing networks on Kropotkin, the first person I must heartily thank is Felix Driver for our inspiring discussions on this topic. On the topics of Reclus and Kropotkin, the British Isles, historical geography and print cultures, I had outstandingly fruitful conversations with Mike Heffernan, Gerry Kearns, Charles Withers, Veronica Della Dora, Heike Jöns, Dave Featherstone, Innes Keighren, Miles Ogborn, Kent Mathewson, John Clark, Petros Petsimeris, Béatrice Collignon, Marie-Claire Robic, and the dear departed Anne Buttimer. From all of you, I have learned a great deal. For discussions on anarchy and geography, I acknowledge especially Simon Springer, Richard White, James Sidaway, Anthony Ince, Marcelo Lopes de Souza, Ruth Kinna, Carl Levy, Matthew Adams, Bert Altena, Constance Bantman, Pietro Di Paola, Ole Birk Laursen, Lucien Van der Walt, Marcella Schmidt di Friedberg, Fabrizio Eva and all the members of the Italian Anarchist Federation, in particular Gianandrea Ferrari, Giorgio Sacchetti and Massimo Ortalli. For collections of *Freedom* and discussions on anarchism and Ireland, I very gratefully acknowledge my friends of the Dundrum House Pub, who are also outstanding scholars, Davide Turcato and Pepe Gutierrez. For the 'Scottish connection', many thanks to Pierre Chabard and Alexandre Gillet. Thanks also to Peter Martin, for the suggestion to trace Nansen's anarchist connections. Finally, this book was made possible thanks to the kindness of the editors of the Routledge Historical Geography Series, Laura Cameron and Simon Naylor, the anonymous reviewers who assessed the draft of the book providing useful feedback, and the editorial assistance of Faye Leerink and Ruth Anderson. All possible errors remain of course my exclusive responsibility.

Notes

1 Ferretti, *Élisée Reclus*; Pelletier, *Géographie et anarchie*; Springer, *The Anarchist Roots.*
2 Souza, White and Springer, *Theories of Resistance;* Springer, Barker, Brown, Ince and Pickerill, 'Reanimating anarchist geographies'.
3 Springer, 'The Limits to Marx'; Harvey, 'Listen, anarchist!'.
4 Pelletier, *Géographie et anarchie.*
5 Siegrist, 'Historicising anarchist geography'.
6 Kearns, 'The political pivot of geography'; Kearns, *Geopolitics and Empire.*
7 Livingstone, *The Geographical Tradition.*
8 Blunt and Wills, *Dissident Geographies.*
9 Keighren, 'History and philosophy of geography II'.
10 Stoddart, *On Geography*, 128.
11 Davies, 'Exile'; Craggs and Wintle, *Cultures of Decolonisation*; McGregor, 'Locating exile'.
12 Featherstone, *Solidarity*; Featherstone, 'Black Internationalism'; Griffin, 'Making usable pasts'.
13 Agulhon, *La sociabilité.*
14 Ferretti, *Élisée Reclus*; Pelletier, *Géographie et anarchie.*
15 Livingstone, 'Science, text and space'.
16 Ferretti, 'The correspondence'.
17 Dunbar, *Élisée Reclus*; Fleming, *The Geography of Freedom*; Clark and Martin, *Anarchy, Geography, Modernity.*
18 Adams, *Kropotkin*; Avakumović and Woodcock, *The Anarchist Prince;* Miller, *Kropotkin*; Cahm, *Kropotkin*; Kinna, *Kropotkin*; MacLaughlin, *Kropotkin*; Morris, *The Anarchist Geographer.*
19 Bantman, 'Jean Grave'; Bantman and Altena, *Reassessing*; Turcato, 'Italian anarchism'.
20 Bantman, *The French Anarchists*; Di Paola, *The Knights Errant.*
21 Jöns, Meusburger and Heffernan, *Mobilities of Knowledge*; Naylor, 'Historical geography'; Secord, 'Knowledge in transit'.
22 Withers, 'History and philosophy of geography', 69.

23   Baigent, 'The geography of biography', 542.
24   Keighren 'History and philosophy of geography I', 642.
25   Ferretti, 'Publishing anarchism'.
26   Keighren, Withers and Bell, *Travels into Print*; Keighren, *Bringing Geography*; Ogborn, *Indian Ink*; Mayhew, 'Materialistic hermeneutics'.
27   Bond, 'Plagiarists, enthusiasts and periodical geography'.
28   Ferretti, 'Publishing anarchism'.
29   Farinelli, *I segni del mondo*; Minca, 'Humboldt's compromise'.
30   Kinna, *Kropotkin*; McKay, 'Kropotkin'.
31   Cho, Crenshaw and McCall, 'Toward a field'.
32   Hirsch and Van der Walt, *Anarchism and Syndicalism*; Shaffer, 'Latin lines'; Maxwell and Craib, *No Gods no Masters*.
33   Ferretti, 'They have the right'; Ferretti, 'Arcangelo Ghisleri'; Ferretti, 'The murderous civilisation'.
34   Baring, 'Ideas on the move', 569.
35   Harris, 'Archival fieldwork'; Withers, 'Constructing'; Ashmore, Craggs and Neate, 'Working-with'.
36   Bond, 'Enlightenment geography', 66.
37   Griffin, 'Making usable pasts', 2.
38   Withers, 'Towards a history of geography'.
39   Latour, *Science in Action*.
40   Haraway, *Simians*; Raghuram and Madge, 'Towards a method'.
41   Lorimer and Philo, 'Disorderly archives'.

1 The Reclus brothers

Translating science and radical politics in the age of empire

This chapter investigates the early links between the 'British world' and the Reclus brothers, Élisée (1830–1905) and Élie (1827–1904), who both came to London in 1852 following the coup of Louis-Napoléon Bonaparte. Drawing on primary sources, I consider that the impact of this experience was crucial for the biographies of both men, and in many ways. First, the Recluses discovered the 'social question' through witnessing the pauperism of the British working classes and through making the acquaintance of socialist leaders, especially other French exiles, when living in London. Furthermore, Élisée witnessed the effects of the Great Famine during his sojourn in Ireland, an experience which informed forever his views on internal and external colonialism and his anti-Malthusian ideas. Second, they were both impressed by British debates concerning the natural sciences and evolution, which led them to adopt evolution as a key concept for both scholarly and activist work. Finally, Élie, less well known than Élisée, returned to live in London for a few years at the end of the 1870s, collaborated with Victorian anthropologists such as John Lubbock (1834–1913), and developed original views on ethnography, which anticipated the later anarchist anthropologists' appraisal of stateless institutions of so-called 'primitive cultures'.

Despite accounts of political repression not being extraneous to British social and political history as magisterially exposed by E.P. Thompson in his work on the formation of the English working class,[1] historians of socialism have observed that, 'by the early 19th century, [Britain] was well established as a haven for political outcasts. The press was relatively freer and the police institutions less restrictive'.[2] This was certainly the case for refugees from autocratic regimes such as the Russian Empire, where social inequalities and conditions of political and religious liberty were far worse than in Britain.[3] This was also the case for those exiles from the continental revolutions of 1848–1849, events which scholars such as Kropotkin considered as founding moments of the anarchist movement. London was then the place of refuge for some of the most famous European revolutionaries (socialists, republicans or nationalists), such as Giuseppe Mazzini (1805–1872) and Alexander Herzen (1812–1870). At the end of the French Second Republic (1848–1852), a sizeable number of the French socialist intelligentsia including Pierre Lerroux

(1797–1871) and Louis Blanc (1811–1882) found refuge across the Channel, increasing even further in number the community of political refugees. When the Reclus brothers felt the same need to depart France, London was also their first choice because the 1848 revolutionaries were remembered as heroes of their youth, and London was considered as the most likely place to go where one could consider emulating their exploits and possibly meeting them.

A further element of context is worthy of consideration. Thompson highlighted the role played by traditions of religious dissent in awakening 'the dormant seeds'[4] of British radicalism. The Recluses had a similar story of religious unorthodoxy in their family's traditions. The anecdotal evidence I discuss on p. 17 concerning how some people in London believed that the Recluses had been persecuted in France not as socialists, but as protestants, reveals a new element for understanding the multiple reasons for their welcome in different British milieus, considering that London already had a tradition of sheltering French persecuted protestants in the seventeenth and eighteenth centuries.[5] Significantly, in his last appearance at the Royal Geographical Society (RGS) in 1903, Élisée Reclus asserted his Huguenot roots,[6] and one of his most important and life-long contacts in London was Richard Heath, a Christian socialist who claimed a link with the tradition of religious dissent, as I explain on p. 16 and in Chapter 5.

Exiles in the British Islands: discovering social and colonial questions

The Reclus brothers were born to a Calvinist family from south-western France. Their father was a minister for an independent and strict church, which had refused to accept the allegiance given by the official French Protestant Church to the state. The two brothers underwent their first international experiences at a very young age, when they were sent to do their studies in Neuwied, in Germany, in a college run by the Moravian Brothers.[7] Despite the internationalist claims of this institution, the two young French pupils were ridiculed by their fellow pupils of different nationalities, as France was generally despised in all the European conservative milieus in the Restoration period following defeat at Waterloo in 1815. They therefore learnt several European languages alongside having to endure insults and mistreatment for being French, experiencing the problems of daily nationalist-based odium. According to Louise Dumesnil-Reclus (1839–1917), Élisée's and Élie's sister and editor of Élisée's correspondence, the other students 'Hated them as French: *damned Frenchmen, French frogs* or *Froggies, die französischen Schweinigel* [French pigs], everything learnt together with occasional kicks and punches'.[8] According to historian Max Nettlau (1865–1944), known as the 'Herodotus of Anarchism' and one of *Freedom*'s contributors in London, 'young and rich Englishmen'[9] had participated in this bullying.

On reaching adult age, the Reclus brothers declined becoming church ministers, in favour of joining the republican and socialist movements that were in ascendency all over Europe in 1848, and progressively embraced atheism.[10] In 1851, Élie obtained his Theology degree at the University of Strasbourg, while Élisée started geographical studies with Carl Ritter in Berlin, to which he was not permitted to return the following year due to his political views.[11] Both were staying at their parent's home in Orthez during the coup of 4 December 1851 when the French republican period (*Deuxième République*) that had opened with the riots of February 1848, ended. The two brothers tried unsuccessfully to organise republican resistance in their region to the coup. Although not formally investigated for this activity, Élie and Élisée preferred to seek refuge abroad, starting voluntarily on one of the many prolonged periods of exile which characterised their respective lives. On 1 January 1852, they arrived in London having already gained important experience in travel, study and politics.

The first common challenge they experienced was the extreme poverty they had to face in the British capital. The limited correspondence from 1852 that survives, namely some letters that Élisée wrote to Élie, restated in an almost obsessive way the daily problems and anguish which the lack of money entailed. Élie quickly obtained a job as a private teacher for the children of one Lady Sparrow in Huntingdon; in that period, Élisée wrote to him from inner London, Tichborne Street (now Glasshouse Street, not far from Piccadilly Circus) that, while listening to a public socialist speech, he had spent his 'last penny. The day before yesterday, I gave my last half-penny to a poor woman ... Poor like Job, I await your return'.[12] In these letters, material circumstances were always a pretext for moral considerations and literary exercises, as can be seen with Élisée's description of the British family of their friend and former Neuwied schoolmate Richard Mannering: 'Mannering cannot lend anything to me; without my asking, he tried to borrow 5 guineas from his father, but not only did his papa refuse, he also treated me like a brigand: "My son, please be wary of intelligent men without a penny: they are far more dangerous than simple miserable souls"'.[13] Among Élisée's complaints, he mentions the limited size of his room and the cold: 'The fire is out, and I am shivering: how could I do anything useful?'.[14]

Throughout the nineteenth century, London was a hub for the politically persecuted from all over Europe. Among the French, this included opponents of the Second Empire in those years, with exiles from the Paris Commune after 1871, and anarchists during the strongest wave of repression in the 1890s.[15] Historians have recently studied the experience of transnational exile as a constitutive feature of anarchist communities in London, such as the Yiddish-speaking Jews from Eastern Europe, who were mainly occupied in the East End clothing industry, and the Italians.[16] Nevertheless, the hardship of exile was determined not only by political persecution, but also by cultural and linguistic barriers which Louise Dumesnil evoked in quasi-nationalistic terms: '[The Reclus brothers] lacked only the indispensable things: gloves, suits, glittering hats. One could not imagine how the English professed then ... a fervent religion for all these exterior attributes of civilisation ... the inquiring eyes of the hotelier examined rigorously the attire for all these extraneous things, especially of those coming from France, that "country of corruption and profane dissolution"'.[17] It appeared that, coming from France, all French people were therefore suspected of socialism, atheism and sedition. Recalling an episode where John Stuart Mill had refused to receive Pierre Lerroux, Louise Reclus explained 'how *The Times* was proud of the superiority of British methods in dealing with political refugees in comparison with the continental solutions: was it not better to let them starve among public contempt, rather than placing them in a jail where they would become heroes or martyrs?'.[18]

Yet, the exiles' networks provided initial support for the two French boys. In their steamboat to London, they met a fellow revolutionary from Hungary 'who would find for them a furnished room and a decent neighbourhood'.[19] In London, they had their first opportunities to meet personally eminent members of the French socialist intelligentsia such as Louis Blanc, the famous reformist socialist, and Pierre Leroux, a Saint-Simonianist who exerted a certain influence on the Recluses' early thought. The exiles' networks facilitated the provision of shelter and of social relations. The two brothers lived for a while in the boarding house of another French exile, Edouard L'Herminez (1804–1882). Political dissident and protestant minister while in France, L'Herminez was disliked by Élisée according to Christophe Brun.[20] Nonetheless, one of his daughters, Fanny, then aged 13, would become Élisée's second wife in 1869, following the death of his first wife, Clarisse Brian.

From his first early time in London, Élisée Reclus also gained a life-long friend, Richard Heath (1831–1912), a figure who also played a key role in settling Kropotkin in England, as I explain in Chapter 5. Heath was Élisée's virtual equal in age. His willingness to learn French provided one of Élisée's first sources of revenue, as Élisée offered to give lessons for 5 shillings per hour. After receiving Heath's request to learn French, Reclus wrote to his student: 'We will try to study French literature together, my language's etymologies and its relations with English, as it seems to me that you are already competent on rules and principles'.[21] It is worth noting that Reclus wrote this letter in French, justifying this choice: 'I could have written in English, but I would have made some errors, so I preferred writing to you in my mother tongue'.[22] Ironically, Heath would later serve as Reclus's and Kropotkin's copyeditor for their publications in English. In a letter sent to Louise Dumesnil in 1908, after Élisée's death, Heath considered those French classes as an important part of his formation as a young man: '[Élisée's] classes were real teaching, because the transfer of knowledge was led by love for justice and truth. I took these classes only for a short time, but their impression was indelible'.[23] Towards the end of his life, Heath was led to reflect on how much he owed to his meeting long before 'with the marvellously inspired personality of Élisée Reclus ... a veritable master, the first on pedagogical aspects, I had in my life'.[24]

According to Nettlau, Heath's political ideas were a far cry from those of Reclus, although the two men shared a strong interest in the 1848 revolutionaries. For Nettlau, 'Heath always remained under the spiritual influence of the authoritarian [side of the movement, represented by] Mazzini, Carlyle and Quinet; he was fascinated by the collective movements of religious fanatics and oppressed peasants, starting from the Anabaptists'.[25] As I discuss in Chapter 5, the links between Heath and the Recluses also lay in their common interest for a 'humane science'. Heath published his obituary of Élisée Reclus in 1905 in the *Humane Review*, which was later reprinted in a book on the Reclus brothers edited by Joseph Ishill in 1927. In that text can be found one of the clearest testimonies of the difficult material conditions that the Reclus brothers experienced in London, as derived from the stories which Reclus recounted to Heath.

> Sometimes he spoke of the hardships endured by his fellow-exiles, and I pictured them with no other bed save the seats in the parks; but he never spoke of what he himself was suffering. It was only some thirty years later, when I lived in Paris and in constant intimacy with his brother, Élie, that I learnt something of what they went through in London. The lodging that he and Élisée occupied was the smallest imaginable – a mere dressing-room over the doorlight. But there must have come times when they had not even such a refuge as this, for Élie spoke of walking about at night trying to find a sheltered corner on one of the bridges.[26]

In recollections he sent to Kropotkin, Heath recalled the impression Élisée had made on him, as having the feeling, 'that he had had revolutionary experiences',[27] although Reclus never took part in barricade construction or skirmishes in the 1848–1849 period.

It is also ironic that sometimes the solidarity which the Reclus brothers benefitted from in Britain was not always due to their political ideas, but to their family origins. An anecdote told by Gary Dunbar shows how Élie Reclus found an occupation in the Fairfield family, the grandparents of famous writer Rebecca West (1892–1983). In West's recollections, her grandmother Arabella Fairfield, 'had engaged a young Frenchman as tutor to her sons, under the mistaken belief that Napoleon III had exiled him from France because he was a Protestant persecuted by the wicked Catholic Church'.[28] West's recollections also reveal a common mistake of many authors and library catalogues, namely, a confusion between the two Reclus brothers Élie and Élisée: 'My father [George Fairfield] and his brother had been brought up by a French tutor who was a remarkable man: he was Élisée Reclus and he was one of two French brothers who were very famous'.[29] The two brothers' correspondence does not leave any doubt of the fact that the Reclus appointed at the Fairfield house was Élie. However, Élisée was also acquainted with the four Fairfield brothers. According to Rebecca: 'All the Fairfield boys adored the Reclus brothers, and Edward and my father called on them from time to time on the Continent'.[30] Indeed, Arthur Fairfield was acknowledged at the end of the fourth volume of Reclus's *NGU*, in a chapter on Britain,[31] which reveals the lasting significance of those first early years in London for the successive careers of the Reclus brothers.

Reclus in Ireland: discovering colonialism and landlordism[32]

Critically, considering the relative paucity of primary sources on this period, West's recollections as quoted by Dunbar provide some important information on the Reclus brothers' lives. First, it seems that Élie's teaching activities exerted some long-lasting influence on the Fairfields. According to West: 'Reclus never intruded his beliefs on my father and his brothers, but he gave them a lasting interest in ideas, and particularly those which related to political science ... I cannot remember a time when I had not a rough idea of what was meant by capitalism, socialism, individualism, anarchism, liberalism and conservatism'.[33] Second, this relationship helps to explain Élisée's important Irish connection, as the Anglo-Irish Fairfield family brought Élie to Dublin, from where the elder brother recommended Élisée undertake work in the countryside for an absentee landlord called Webster. As Louise stated, Élisée 'did not like at all teaching syntax and was repelled by pedagogical rules. At that time, he felt that he was attracted by agriculture'.[34] This is confirmed by Heath, who recounted to Kropotkin some details on their acquaintance at that time. 'The last time I saw [Élisée] we were in the Old Reading Room at the British Museum, where he told me that he was about to leave England and with several of his friends he was about to cultivate the land ... He was going out to one of the colonies to found a settlement and another kind of society, where everyone would work the land. This and other things he said amazed me, and I well remember wondering at the extraordinary nature of his ideas and character'.[35] The reason for Heath's amusement was that he found it odd seeing an intellectual prepared to work on the land. Indeed, this choice of work had arisen due to the influence of Saint-Simonian socialism on the young Reclus, and inspired a willingness to build utopian colonies for sheltering other political persecuted figures and exiles. As shown by recent scholarship, Reclus progressively abandoned ideas involving the founding of Icarian communities and argued for struggle within the capitalist society in order to transform it.[36]

In Ireland, Élisée obtained a post as an administrator in Kippure Park. Situated in the county of Wicklow, 50 kilometres south of Dublin, this estate comprised 82 hectares of bogland and was owned by a landlord, Webster, based in London. Élisée's letters to Élie show that this Irish experience was a very formative one not only for the future geographer, but also for the future anarchist. At the time, Ireland was suffering the direct consequences of the Great Famine, which Reclus addressed in his *NGU*, having been deeply affected by the misery and depopulation of the Irish countryside. In 1852, Reclus had plans to transform the land to use it more productively using recent technological advances, and to drain the bogland, and asked Webster to invest the necessary capital, because in his mind 'agriculture here has remained in the state it was in perhaps in the days of the ancient Celts'.[37] Disagreements with Webster over these modernising proposals led to Reclus' departure and bitter disappointment. The geographer wrote at the time to his brother that: 'The property is in disarray, and dissatisfaction has reached dreadful proportions'.[38] During a trip to western Ireland, Reclus visited another of Webster's employees, named Pennefeather, an Irishman whom the French exile mocked for his unconditional submission to all authorities, relating the gesture with which Pennefeather accompanied the phrase (quoted in the original), 'I delight in mi queen, in mi noblemen, in mi gentlemen'.[39]

In Wicklow, Reclus developed the idea of settling in an isolated place to cultivate land with other French exiles, a proposal which he subsequently attempted to put into practice some years later in Colombia.[40] Celtic traditions also exercised a fascination on his own geographical imagination. In *L'Homme et la Terre*, Reclus described Ireland as one of the most civilised countries in Europe in the early Middle Ages, because 'Ireland had escaped Roman conquest: the peoples of Erin had never been broken and degraded by servitude like the Gauls and Bretons, so they have maintained more initiative and strength, as well as a greater freedom than other Christians in their way of believing'.[41] Consistent with his anticlericalism, Reclus went on to note the responsibility of Pope Adrian IV for the English invasion of Ireland in the twelfth century, stressing the ironic contradiction between the affection of many Irish for the Roman Catholic Church and the historical faults of the latter, which had been implicated in establishing the island's colonial condition:

Henry II built his palace in the city of Dublin to establish forever his power, so the Irish were deprived of their independence and of their own civilisation; thrown into poverty and barbarity, they started the painful phase of their history characterised by servitude and degradation, which has continued to this day and, by a strange irony of fate, has closely linked them to this Roman Church by which they had been sold to England.[42]

According to Reclus, the later involvement of many Irishmen in Europe's religious wars in the sixteenth century, resulted in the fact that, 'doubly enemies to the English, they were then doubly oppressed, firstly as Irish and then as Catholics'.[43]

In Reclus' anticolonial critique, a loss of freedom and independence is always presented as a cause of backwardness and 'moral degradation' for the peoples concerned, who come under an ethical obligation to revolt against oppression. In this sense, Reclus' arguments are akin to those advanced a century later by Edward Said (1935–2003), who reasoned that the place of colonialism in Irish history has to be studied to understand contemporary Irish identities. According to Said, failing to see Ireland as a postcolonial country would mean that all its problems 'are its own and certainly cannot be ascribed to British colonialism'.[44] It is thus possible to conclude that Reclus' sojourn in Ireland was critical in shaping his ideas on social justice, as he directly witnessed extreme poverty and hunger, and developed a critique of 'voluntary servitude', expressed in his severe ridicule of the psychological enslavement he found in people he met in respect of the Crown, the Church and the landlords.

In autumn 1852, Élisée Reclus decided to seek land further away to fulfil his utopian dreams and left for the United States, embarking from Liverpool in December 1852. Thus began a five-year period in which he resided in Louisiana and in Colombia, acquiring an awareness of slavery, racism and the ongoing challenges of trying to establish utopian communities, before returning to France in 1857.[45] Meanwhile, Élie remained in the British Isles until returning to France in 1855.[46]

Long-lasting effects of the London experience: dealing with British science[47]

The experience of living in London left an indelible mark on the two brothers. In Élisée's case, it fostered his knowledge of new geographical works, such as the Great Globe by James Wyld (1812–1887), built in Leicester Square for the 1851 London exposition, and which inspired his own project involving a Great Globe for the Paris Exposition of 1900 that was intended to stretch 127.5 meters in diameter.[48] In London in early 1852, Élisée made his first attempt to write a geographical essay, though it was quickly aborted. As he wrote to Élie on 2 March 1852: 'My *Japan*, a definitive failure, was badly conceived and more difficult than what I had imagined'.[49] London was a very important place for the scientific training of both brothers, because they started in London to integrate their initial literary and theological education with the study of British natural sciences, on the eve of the Darwinian revolution. This was confirmed in 1858, once they were both back in France, in a letter to Auguste Nefftzer, the director of the newly founded journal *Revue Germanique*, which would be one of Élie's main publishers in the following ten years:

> We offer our services to the *Revue Germanique*, whose programme immediately acquired our sympathies, to the limit of our capacity. We are two brothers who lived in Germany for a long time and in different periods, so that the German language has no difficulties for us. We both studied protestant theology, but we refused to exercise the Ministry for reasons that free thinkers will easily appreciate. Philosophically, we adhere to the school of Spinoza.[50]

Élie and Élisée indicated their willingness to translate 'Oken, Schubert, Baader, Steffens, Goerres, Hamann' and offered 'a short paper on Feuerbach's *Theogony*, where we have tried to explain his system in our own words', explaining that, 'with our theological studies over, we devoted ourselves to the natural sciences in England'.[51] This letter is significant for understanding the

Recluses' role in cultural transferral and translation within the scholarly and intellectual currents circulating in different European languages. Following studies in Germany, their stay in the British Isles in the 1850s was instrumental in allowing them to assimilate new scientific theories circulating in British debates at that time. It also accounts for the political nature of their subsequent writing in various fields.

The commitment of early anarchist geographers to evolutionary theory is impossible to understand without considering the importance of 'rational' science as a battlefield between religious traditions and liberal, secular and progressive free thinking. Recent works by James Secord on Charles Lyell (1797–1875) are an important starting point in clarifying this topic, in relation to the interest of early British atheists and free-thinkers in the *Principles of Geology*. [52] In fact, the physical sciences were involved early in questioning biblical stories such as the Flood, and were clearly at the centre of a fierce controversy whose stakes involved 'freeing science from Moses'.[53] In this context, geology was deemed 'an upstart science associated with infidelity and revolutionary atheism'.[54] Secord contends that Lyell became one of the main inspirations of the later Darwinian revolution, and that his science can be seen as the expression of an implicit political strategy by someone who was not a revolutionary, but a gentleman with important ties among the upper class. Thus, Lyell's book 'was a Trojan horse' in a very real way, as only a gentleman 'could have used such an unlikely vehicle to advocate controversial views. The *Principles* had the imprint of conservative classicism, but had within a secret army of reform'.[55] In the 1860s, the Recluses were also among the first French readers and translators of John Tyndall,[56] whose celebrated 1873 address in Belfast has been recently studied by David Livingstone. Livingstone considers that the address marked a key moment in understanding the stakes involved concerning the well-known controversy between British scientists and the clergy, one which Livingstone understands more as a dispute for gaining professional advantage rather than as an ideological dispute prompting atheism,[57] as it would be in the case of the anarchist geographers.

According to George Stocking, some British anthropologists used Darwin's ideas against most conservative and clerical thinkers through declaring, as a principle, the fundamental unity of the human faculties.[58] This approach implied the inclusion of so-called 'savages' in a common human story, challenging conservative authors such as Joseph de Maistre (1753–1821) who, in espousing the concept of 'degenerationism', rejected the very possibility of progress and considered 'savages' as being not humans but a degenerated branch of Adam's lineage, the 'objectification of original sin'.[59] In a clear contrast with evolutionary anthropologists, conservatives and clerics declared the 'innate' inferiority of 'primitive peoples' in the political context of the Restoration. According to Ugo Fabietti, 'the savage, the Other, found a place in the history of humankind through the optimistic and progressive ideology of the Enlightenment: after the latter's decline, [the savage] was banished once again from history'.[60] Evolutionary anthropologist John Lubbock then questioned 'not only the biblical tradition, but also the chronology of the world that was accepted by the Church',[61] introducing the concept of pre-history. The crisis of belief in creationism and diluvialism was accompanied by what Fabietti calls the 'Palaeolithic equation', which refers to the belief that all humankind was likely to have passed through the same developmental phases of material culture (including, for example, the Palaeolithic, Neolithic, and Metal ages), and thus 'primitive peoples' were not 'degenerated' people, but an image of early Europeans, 'due to the fundamental identity of human faculties'.[62] Recent work on Lubbock has shown a widespread awareness of his 'Gospel' among the British middle classes,[63] the same social categories that would constitute the main public for Kropotkin's articles on evolution and mutual aid.

This antagonism between science and religion, and between ethnocentrism and attempts to understand different peoples, was clearly an important motivation for the early anarchist geographers to engage with evolutionary science. Elements of continuity between biology and sociology which existed in the work of many British evolutionary theorists were not unfamiliar to the Reclus brothers, given their previous experience of German intellectual milieus steeped in Lorenz Oken's and Friedrich Schelling's *Naturphilosophie*. The ideas of these German thinkers had inspired geographers such as Humboldt and Ritter, and considered humankind and the environment, nature and culture, history and geography, not as separate realms but as related entities.[64] This refusal to separate nature and mind would equally inspire Kropotkin's work on mutual aid, an understanding he applied in relation to plants, animals and humans.[65] For Élie Reclus, London became significant again in the aftermath of the 1871 Paris Commune. Like his brother, Élie found refuge in Switzerland, without ever losing his contact with British scholars and intellectual institutions. A place of refuge for many Communards, London continued to exert an attraction for intellectual exiles, especially in terms of the possibilities it offered for publishing, for attending learned societies and for finding materials in libraries and archives. During his time in Zurich from 1871 to 1877, Élie Reclus developed an increasing interest in the fields of ethnography and in the study of popular beliefs and traditions. A reader of Scottish anthropologist John Ferguson McLellan (1827–1881) and in correspondence with Swiss scholar Johann Jakob Bachofen (1815–1887), Élie began studies on matrilineal kinship, which was a topic closely connected with the Recluses' critique of formal marriage and their collaboration with early feminist movements, as shown in recent work.[66] In 1877, Élie wrote a long review of Bachofen's *Mutterrecht* and McLellan's *Primitive Marriage* in the *Radical Review* edited by North American anarchist Benjamin Tucker, which demonstrates a persisting link between scholarly and activist concerns in the choice of his research topics.[67]

In 1875, Élie made contact with Lubbock in London in a letter written from Zurich requesting Lubbock's patronage to assist with 'entering this scientific field', a description Élie further clarified evoking his 'definitive preference for studies of ethnology ... I am interested in a number of subjects addressed by your Society ... Therefore, Sir, if my first work is accepted, the Anthropological Institute will soon have the second'.[68] The Anthropological Institute of Great Britain and Ireland had been established in 1871 following the amalgamation of the Ethnological Society and the Anthropological Society, which had brought together different interests and scholarly tendencies, from physical anthropology to studies on folk-lore and popular beliefs, from polygenists to monogenists.[69] This intellectual pluralism reflected the views of the elder Reclus, given his interest in ethnographical studies on so-called 'primitive peoples' as well as in comparative histories of myths and religions. At the end of 1876, Élie lost his principal source of revenue, which had consisted in a longstanding relationship with the Russian liberal journal *Delo*, following a hardening of political censorship in Russia at the beginning of the Russian-Turkish war. He then looked for new work, and was given the opportunity to be involved in publishing work in New York. According to Élisée's recollections, Élie then,

> left for the United States, where he had been invited to collaborate with one of the principal reviews. He was of course courteously received, but after the first article, there was a dispute: had not Élie had the shamelessness to speak of the Brothers Goncourt and their work, the *Fille Elisa*. The editor of the American review declared that the moral purity of his readers interdicted to him to treat of such a question; in consequence, Élie, after a journey of study, and the gathering of impressions in the States of New York and Massachusetts, left for hospitable England, where he had passed the years of his first exile.[70]

The description 'hospitable England' leaves no doubts; despite all previous difficulties there, London as an international focal point was perceived as a preferred destination, as it remained for other unorthodox intellectuals and political dissidents from the Continent, including of course Kropotkin. Among Élie's new acquaintances in London there included the sub-editor of *Nature* and future secretary of the RGS, John Scott Keltie, as evidenced in Keltie's unpublished letters to Kropotkin: 'I had a long talk the other day with your friend Élie Reclus'.[71] Nonetheless, Élie's sociability networks were not limited to scholars; according to Christophe Brun, Élie's 'salon' in London was visited by several political dissidents including the German 1848 exile and former friend of the Recluses 'Eugene Oswald, and [by] the socialist feminist Jeanne Deroin'.[72]

Élie was received as a member of the Anthropological Institute on 27 November 1877, and remained an active member of the Institute until his return to Paris in late 1879. In 1878, he presented to the Institute a paper titled 'On Circumcision: its significance, its origin, and its kindred rites'.[73] However, the most important work in relation to Élie's early anthropological writings appeared in two of the most famous publications of that time. The first was the London *Cornhill Magazine,* then edited by Murray Smith and William Thackeray, where Élie Reclus published a paper on the 'Evil Eye'.[74] In the following years, Kropotkin also collaborated with this journal, and corresponded with the new editor Reginald G. Smith.[75] The second was the ninth edition of the *Encyclopaedia Britannica,* considered as 'the scholars' edition, then edited by Thomas Spencer Baynes (1823–1887), where Élie published a chapter on 'Ethnography and Ethnology'[76] in volume eight, and on 'Fire' in volume nine.[77] The fact that Élie was one of few Frenchmen invited to contribute to this work, an honour also extended to Kropotkin but not to Élie's famous brother Élisée, indicated the prestige Élie Reclus enjoyed in British scholarly milieus in the late 1870s.

At that time, 'ethnography' was a relatively new subject area,[78] and Élie Reclus was one of the first to give an authoritative interpretation of this term in both French[79] and British publications. In the *Encyclopaedia Britannica,* Élie asserted that: 'Ethnographists deal with particular tribes, and with particular institutions and particular customs prevailing among the several peoples of the world, and especially among so-called savages. Ethnologists bring simultaneously under review superstitions, legends, customs, and institutions which, though scattered in distant regions of the earth, have some common basis or significance'.[80] At the same time, these definitions complemented the unification of different areas of investigation promoted by the Anthropological Institute, and pointed towards the future commitment of Élie Reclus as an 'advocate' of so-called 'primitive peoples', and as a historian of religions later at the University of Brussels, where he was joined by his brother Élisée in 1894.[81]

Reclus in London on numerous occasions

Unlike Élie, Élisée never went to live in London again long-term, but he often travelled there for short though very active periods. In 1862, London hosted the Universal Exposition. Élisée Reclus was then a member of the editorial body of the Guides Joanne, a French series of best-selling travel guides produced by the major publisher Hachette, the principal French employer of the Reclus brothers.[82] In 1862 Élisée, already the author of a traveller's guide to London and its environs,[83] travelled twice to London to complete an illustrated guidebook on the Universal Exposition.[84]

The first trip took place in February 1862, and is only documented with a few lines in the personal diary of Reclus's first wife, Clarisse Brian (1832–1869).[85] A longer stay occurred between August and September of the same year, documented in Élisée's correspondence. The anarchist geographer travelled with a group of colleagues and friends including a former colleague from his days of Protestant studies, Gustave Hickel (1821–1870), and a member of the Paris Geographical Society and Hachette contributor Ernest Morin.[86] The reasons for Reclus's trip were both professional and political. The first anecdote he told Élie in his letters from London, reminiscent of their earlier experiences there together, concerned his group's decision 'to quit a hotel where a Frenchman, looking like a police spy, investigated us under the pretext of being a fellow citizen. [Instead, we] settled at the house of a very kind English lady'.[87] During this visit, the first stop-over occurred at Crystal Palace, freshly refurbished for the upcoming Exposition, the notes of which demonstrate Élisée's literary talent. That talent indicates one more reason his published works enjoyed such success among the general public. However, these visits also represented an opportunity to obtain updates on scholarly matters, by gathering information not available elsewhere. Visiting the Exposition, Reclus was struck by the standing of Austrian scientists whose works were displayed with a strong emphasis on geographical and geological sciences. According to Reclus: 'Formerly, I had no idea of such a scientific movement in Austria. In comparison, all the other nations risk looking unproductive in the great movement of intellectual work'.[88] London remained, therefore, a principal destination to become aware of and gather resources for intellectual work.

In terms of politics, London workers had just gathered to protest against czarist repression in Poland, and Reclus strongly commended these social movements 'where an embryonic International can be seen'.[89] The First International, a shortening of its official name, the International Workingmen Association (IWA), was founded in London in 1864. Acknowledged by Kropotkin as one of the milestones of anarchist history,[90] this event swiftly had consequences for the Reclus brothers, who soon joined the Parisian section of the IWA and made the acquaintance of Russian revolutionary Mikhail Aleksandrovic Bakunin (1816–1876), who was a major figure alongside them in the definitive rupture between anarchists and Marxists in the International, officially confirmed in 1872 at the Congress of Saint-Imier.[91]

In the summer of 1869, Élisée travelled to London again, invited by the General Council of the IWA, bringing with him his nephew Paul, Élie's son, then 11 years old and later a refugee in Britain from 1893 to 1903.[92] Reclus's account of the first event he attended, a public meeting of republican secularists, reveals two characteristics of his political engagement. First, his willingness to engage in a wide-ranging inclusiveness (which he was to reconsider after the Paris Commune) of different leftist political positions; and second, his endorsement of women's political participation: 'We were 120 men and women, all together, free-thinkers, republicans and socialists, and happy to be united'.[93] Reclus's assessment of British radical milieus was multi-faceted. On the one hand, he stressed the difficulties in cultural communication between the British Isles and the continent. 'England is surely changing for the better but, excluding the world of workers, she is very ignorant concerning continental matters'.[94] Yet, after attending IWA meetings, Élisée stated that: 'The English are more communist than what I had believed. ... Even the nationalization of the telegraph service, as advocated by Gladstone, is interpreted by them in a communistic sense. After the telegraph service, will come the railways, and after transport will come production ... they wish to promote pure communism not only because it is a social necessity, but because it is "the ideal" ... Indeed, they consider communism as a British invention'.[95] It is worth considering that the 'Founding Fathers' of anarchism such as Bakunin, the Recluses and James Guillaume, did not apply to themselves the description of 'communist anarchists' (only adopted in the second half of the 1870s), preferring looser descriptions such as 'anti-authoritarians', 'republicans' or 'social-democrats'. Therefore, some influence from British terminological preferences on the future descriptive label of 'anarchist communism' cannot be excluded, as might also be presumed owing to the friendly relations between Kropotkin and British non-anarchist socialists that I discuss in Chapter 5.

In 1869, recently widowed, Élisée again met Fanny L'Herminez, who was working then as a primary school teacher and she agreed to go with him to Paris to become his wife; they did not marry legally, but celebrated a 'free union' with relatives and friends. This event anticipated the free unions publicly celebrated by Élisée's daughters Magali and Jeannie in 1882 with their respective partners, which caused wide-spread scandal among the French bourgeois press, and puzzled some British scholars such as Scott Keltie.[96] On 4 April 1871, Élisée Reclus was captured during one of the first skirmishes between the Commune's National Guard, to which Reclus was attached, and Versailles troops. The outcome of this episode revealed the special sympathies that Reclus enjoyed in the British scientific world, where he was also becoming popular thanks to the English translation of the first of his substantial geographical works, *La Terre* in two volumes (1867–1868), published in London as *The Earth*, and edited by Henry Woodward that same year.[97]

During a period known as 'Bloody Week' from 21 to 28 May 1871, the Paris Commune was crushed and a harsh repression was applied to those who had fought in its defence.[98] On 15 November 1871, Reclus was sentenced to ten years deportation to New Caledonia, which would have probably resulted in his death given the high mortality rate of people deported to French penal colonies. In France, only his publishers and a few members of the Geographical Society tried political lobbying to help Reclus,[99] while the British scholarly world stood up in substantial numbers in favour of the French geographer, and at least two petitions were addressed by several dozen British activists and scientists to the President of the French Republic Adolphe Thiers, requesting clemency for the scholar. According to Gary Dunbar, 'a persistent story has it that Charles Darwin was among the signers, but his name does not appear in the petition'.[100] However, Dunbar's assertion only relates to copies of the petitions, published by the French press, which survive in Reclus's archives at the French National Library, while, to the best of my knowledge, nobody has found a copy with the original signatures. Darwin's archives show that the British naturalist was aware of Reclus's work, as a few months earlier, in February 1871, he had written to a Dutch correspondent expressing interest in Reclus's travel accounts on South America in relation to evolutionary theory, inviting his interlocutor to inquiry in Paris to know whether 'M. Reclus is trustworthy'.[101] However, my survey of the British press in the 1871–1872 period adds what I consider as a new and decisive element to this discussion, namely, that Darwin's name was mentioned by at least a dozen British newspapers between December 1871 and January 1872 as one of the signatories, and apparently no retraction was published in the following months.[102] This indicates that, while Darwin might have not signed this petition, he was certainly not opposed to having his name associated with it. This further confirms the existence of a mutual feeling of respect between the Recluses and 'British Science', in the context of what British scholars had called 'the commonwealth of letters', a translation of the French Enlightenment's idea of the *République des Lettres*.

The text as published by the *Pall Mall Gazette* read:

> To the President of the French Republic. As foreigners, yet owing a debt of gratitude to France and to her people, some of us for long years of pleasant sojourn, some for equally pleasant journeys, some for life-long friendships, and all for treasures of thought and delight derived from her ample stores of literature, science, and art, we venture to apply to you, Sir, on behalf of a writer of whom we believe France to have reason to be proud, now lying under a heavy sentence. It is not for us here to express any opinion bearing on the internal politics of France. But we dare to think that the life of a man like Monsieur Élisée Reclus—whose already widely acknowledged services to literature and science did but promise, from the ripeness of his vigorous manhood, services more signal still for the future—belongs not only to the country which has given him birth, but to the world, and that in reducing such a man to silence, or sending him a prisoner beyond the pale of civilization, France will be but crippling herself, and diminishing her influence over the world. Leniency, indeed, always becomes victor; but how much more so when the victor is France, and among the vanquished is one of the young celebrities of French literature and science. Surely, sir, your own name is too illustrious, your place too eminent in the commonwealth of letters, to allow Monsieur Elise Reclus' deportation to cast a blot on the literary renown of your great country. With fervent wishes for the prosperity of France.[103]

The petition's promoters included Eugene Oswald (1826–1912), a German refugee of the 1848 revolution and an old friend of the Recluses, and John Malcolm Forbes Ludlow (1821–1911), a supporter of Christian socialism. Among the other signatories, it is worth noting a number of past and future acquaintances of the Recluses and their networks. The first, Henry Fawcett (1833–1884), professor of Political Economy at Cambridge, was the brother-in-law of Elizabeth Garrett who had been the Recluses' host in Paris. The second, John Russell, Viscount Amberley (1842–1876), was a friend and future brother-in-law of Thomas Cobden-Sanderson (1840–1922), who would shelter Paul Reclus in 1894 during his escape from French police.[104] In February 1872, Élisée's sentence was commuted to banishment from France, and he moved to Switzerland. The effectiveness of the British petitions was directly acknowledged by Reclus in a letter sent to Oswald from Zurich on 21 March 1872: 'I owe you my freedom, or at least an anticipation of many months in recovering control of my life'.[105] A new period in Reclus's career was beginning, for it was in Switzerland that the network of the anarchist geographers was first constituted and where the Recluses met Kropotkin.[106]

During Élie's 1878–1879 stay in London, Élisée, still exiled in Switzerland, worked on the volume of the *NGU* dedicated to north-western Europe, which included the British Isles. The elder brother provided first-hand information for Élisée's work, sending him notes on that region which were acknowledged in the final work.[107] Élisée's unpublished correspondence with Hachette publishing official Charles Schiffer shows that, in 1878, the geographer requested his editor's permission to modify the order of volumes. Among the geographer's considerations put forward to support this change, there is a request to travel to work 'at the British Museum, which has many advantages'.[108] Reclus stayed in London in autumn 1878, where he met British scholars such as Ernst Georg Ravenstein (1934–1913), Ludlow and Woodward.[109] As was typically the case for Reclus's style of work, local networks near sources of information were paramount to obtaining first-hand information and pre-publication feedback for the *NGU*.

In another letter to Schiffer, Élisée announced a new visit to London between 1884 and 1885 'to search all documents related to the Niger basin which are held at the [Royal] Geographical Society library'.[110] This journey is not documented in other sources, but its purpose reveals the complex relations between Reclus, Kropotkin and the imperial milieus then gravitating around the RGS, first analysed by Gerry Kearns.[111] On the one hand, the anarchist geographers demonstrated a great interest in the cosmopolitan intellectual life of British imperial circuits; on the other hand, they condemned in the harshest terms the militarism and colonialism of those promoting such policies. This would become even more apparent during the last trips Élisée undertook to London from 1895 to 1903 in association with Kropotkin, when both men focussed on editorial projects with RGS scholars such as Mackinder, Mill and Scott Keltie. In most cases, activist and scholarly interests merged during Reclus' trips, involving both scholarly addresses and speeches to activists. In Reclus's final years, his visits to Britain were facilitated by his appointment to Brussels University, in a city which he chose also because of its greater geographical proximity to the British world than Switzerland.[112]

In February 1895, Reclus travelled to London for several weeks to receive the Patron Medal he had been awarded from the RGS the previous year, 'for having devoted his life to the study of comparative geography'.[113] Concurrently, he had an opportunity to visit Dorset, where he met for the first time the famous naturalist and evolutionary theorist Alfred Russell Wallace (1823–1913), known for his progressive political views, who was also in correspondence with Kropotkin, as I discuss in Chapter 2. This meeting made a significant impression, recalled by both men. Reclus wrote to his sister Louise that he was 'very happy to see this man exuding moral powerfulness and goodness. Here you have someone putting his efforts to good use: his only concern is "Land to the Peasants"'.[114] Wallace's impression was likewise approving. Although he considered Reclus as 'the greatest of geographers', the two men spent the afternoon discussing not geography, but anarchism:

> I was very anxious to ascertain his exact views, which I found were really not very different from my own. We agreed that almost all social evils—all poverty, misery, and crime—were the creation of governments and of bad social systems; and that under a law of absolute justice, involving equality of opportunity and the best training for all, each local community would organize itself for mutual aid, and no great central governments would be needed, except as they grew up from the voluntary association of their parts for general and national purposes... Few would think to look at this frail man that he was not only in the very first rank among the students and writers of the nineteenth century .... He has now passed away (1905), having completed one of the greatest (if not the very greatest) literary works of the past century. But he will also be remembered as a true and noble lover of humanity.[115]

This confirms the strong links existing between anarchism and evolutionary natural sciences, and that many anarchists considered the latter as providing intellectual tools to challenge religions through fostering the development of rational science, as well as regarding cooperation as an evolutionary component facilitating the discrediting of Malthusianism.[116]

In the summer of 1895, Reclus took advantage of a second invitation to the Edinburgh Summer Meetings to pass through London to give public lectures on anarchy, the first of which took place at the South Place Institute on 29 July 1895. The correspondence between Kropotkin and Alfred Marsh reveals the great care with which the occasion was prepared for by the local anarchist group editing the journal *Freedom*. It appears that local activists felt that they were receiving a 'guru' and needed to benefit from the occasion. From Viola Cottage, in Kent, Kropotkin wrote several instructions to Marsh, including that they did not have to introduce the speaker, as 'E.R. has a worldwide reputation'.[117] A multilingual reader, Reclus always considered himself as a bad speaker and was never comfortable with spoken English. Kropotkin therefore informed Marsh that 'Élisée writes out his lectures as he fears that he has not mastered English well enough'; Reclus was willing 'to answer questions' but it was good to have the help of 'an English comrade'[118] for the discussion. These linguistic difficulties were also brought up in a letter from Élisée to his third wife Ermance, mentioning a conference he could give in French, in Edinburgh, as the attendees were intellectuals, all able to understand him. But the following address, in Glasgow, had 'to be done in English, before a public mostly composed of anarchist workers. This will be the most difficult part of the schedule'.[119] This experience confirms recent scholarship on geographies of internationalism, which have contended that scientific life is not straightforwardly international and that the circulation of knowledge between different cultural and linguistic areas needs a strong voluntary commitment among those involved for it to operate effectively.[120]

However, Reclus also considered international scholarly cooperation as a way to foster internationalism, as shown by his campaign for the realisation of a Great Globe for the Paris 1900 World Fair, a project that he presented first at the International Geographical Congress held in London in August 1895, in collaboration with Paul Reclus.[121] While there, Reclus met for the first time a number of British geographers who became his correspondents over the following years, such as Mill and Mackinder. The political significance of Reclus's international commitment is also demonstrated by the names of the foreign guests who were involved in a university extension meeting in Brussels in the following October. As Reclus wrote to his sister Louise: 'English guests, my co-geographers Mackinder and Parkyn made very interesting speeches … it is possible that this meeting will entail some progress for internationalism, with an exchange of lectures and lecturers between Belgium, the Netherlands and England'.[122] University extension programmes were one of the most important places for collaboration between anarchist geographers and more 'mainstream' scholars, given the importance which anarchists attributed to the education of adults belonging to the popular classes, a commitment which was shared by social reformers such as Patrick Geddes.[123] Reclus was also interested in the 1:1,000,000-scale map of the Earth project proposed by German geographer Albrecht Penck, then being discussed in international geographical congresses,[124] considering it as an opportunity to use science to cross cultural and national boundaries. As he wrote to his cartographer Charles Perron, in proposing to associate the Great Globe project with Penck's endeavour: '[Penck] is German, and by our possible alliance we could escape from this dishonourable and awful impasse of "French science and German science"'.[125] Travelling to London also meant helping to internationalise science.

Reclus's following trip to London, in the summer of 1896, seems to have been undertaken mainly for activist purposes. Reclus attended anarchist meetings in preparation for the upcoming International Workers Congress. The most important recorded initiative at that time was the large anarchist meeting held at Holborn Town Hall on 28 July 1896, representing 'the anti-parliamentarian, syndicalist and anarchist tendencies',[126] whose speakers comprised an impressive list of 'big names' of international anarchism, including Reclus, Kropotkin, Christian Cornelissen, Louise Michel, Bernard Lazare, Ferdinand Domela Nieuwenhuis, Errico Malatesta, Pietro Gori, and Gustav Landauer. London was at his apogee as the centre of transnational anarchism at the turn of the nineteenth century. According to a short report published in *Freedom*, Reclus spoke at the Holborn Town Hall Meeting, arguing for the power that an idea can represent against laws and institutions:

'When an idea grows and grows, revolution must come: it is impossible to stop it, it will come, it will come'.[127] In 1898, an English translation of Reclus's celebrated pamphlet 'Evolution, Revolution and the Anarchist Ideal', which developed and clarified his ideas concerning anarchism, was enthusiastically advertised on the first page of *Freedom*.[128]

The following visit to London, in 1898, was mainly dedicated to geographical business. Reclus needed to promote the Great Globe project, and his presentation to the RGS on 27 June 1898 has been considered as the last attempt to obtain international support for this monumental endeavour, which was abandoned that same year for lack of funding.[129] Reclus's letters to Kropotkin show that, already in early 1898, the French geographer lacked confidence regarding the likelihood of the Great Globe project going ahead and 'would have been surprised'[130] if it were to proceed. However, what is significant is that British scholars were generally very supportive. At the RGS, Reclus found scholarly solidarity in contrast to a French public debate which had turned disappointingly in the direction of economic considerations. At the discussion which followed his session, Patrick Geddes, who was actively involved in promoting the Great Globe project with Paul Reclus in Edinburgh, defended Élisée Reclus's position with regard to debates happening in France, arguing that:

> If Mr. Reclus chose to carry out his undertaking on lines of popular amusement and advertisement, it might have been carried out a long time ago. But Mr. Reclus has maintained a strict attitude in this respect. Serious and practical geography demands, therefore, that it is necessary for scientific men to help him to maintain this attitude, and not to vulgarize science by making a vulgar exhibition of it in any way, but to carry out the scheme of such a temple of science as he has so eloquently described.[131]

This position was substantially endorsed by two other participants in the discussion, J.B. Jordan and Henri Brin. The President (Clemens Markham) concluded by proposing 'that a scientific committee be formed to assist and to co-operate with M. Reclus in his great attempt to improve the methods of constructing'.[132] Once again, the British scientific world had supported Reclus when he faced issues in his own country. Although the proposed committee was never formally established, networks of collaboration continued to work across the Channel, especially thanks to Geddes and Paul Reclus.[133]

In his personal capacity, Wallace supported Reclus's Great Globe project, writing an article for the *Contemporary Review*.[134] Although criticising some technical aspects of the project, including its costs and feasibility, Wallace grasped and appreciated one of the most strategic points of Reclus's project: 'Such a globe would correct erroneous ideas as to the comparative size and shape of different regions due to the use of Mercator's or other forms of projection'.[135] This observation corresponded to what today would be called a critique of Euro-centrism in mapping. Wallace's alternative proposal was to deploy Wyld's model and to construct the world's 'true' relief not in terms of the external surface of the globe, but in terms of how it could be viewed internally, to facilitate both visits and creative thinking. Reclus's objections to this solution have been well summarized by Gary Dunbar, who contended that the anarchist geographer 'wanted to remain faithful to Nature'.[136] For Reclus, 'A plane surface never will nor can be a real representation of a spherical surface'.[137] It is worth noting that the anarchist geographer still evoked his youth experience in London as an inspiration. 'I remember, also, the real feeling of rapture which pervaded me when, in my youth, I walked inside of Wyld's globe, near this very place, admiring the magnificent sight'.[138]

Reclus's final visit to London, in April 1903, was organised to present several models inspired by his thinking concerning the Great Globe, namely, the globular maps, or spherical atlases, which were his last attempt to build three-dimensional models of the world, in collaboration with Brussels cartographer Emile Patesson. It was on the occasion of this visit that Reclus pronounced his famous declaration on bi-dimensional maps, which 'ought to be entirely tabooed. They must be tabooed, because maps are made on different scales, and that being so, it is quite impossible to compare them; and if you cannot compare them, it is only a waste of time and trouble. In all well-conducted schools, globes should be used, and children ought to be entirely forbidden the use of maps'.[139] In a subsequent discussion, Oxford geographer Andrew John Herbertson (1865–1915), a former student of Geddes and a Reclus admirer, recalled a previous Reclus visit to Oxford, not documented in other sources, which apparently had taken place in 1901: 'A year and a half ago M. Reclus was good enough to present one of the curved maps to the School of Geography in Oxford, and I have constantly used it'.[140] Though supportive, the assembly questioned Reclus's way of depicting three-dimensional geographical objects, and some of the major authorities of British geography, such as Herbertson, Mackinder and Ravenstein, intervened in support of the possibility of deviating from what Reclus defined as a basic and fundamental postulate, namely, to never exaggerate the vertical scale in relation to the horizontal scale.

In a letter sent to Patesson the day following these events, Reclus revealed both satisfaction and disappointment. His satisfaction was due to the fact that: 'The room was more than full and all the intellectual geography of England was represented, including the professors of Oxford, Cambridge and Manchester'.[141] His disappointment arose because Reclus felt that his ideas on the necessity of constructing reliefs at the same scale vertically and horizontally had not been taken seriously, as even Kropotkin allowed for deviations to his preferred approach for reasons of improved visualisation. Despite this, Reclus hoped that the British market would allow the fabrication of a certain number of spherical atlases through securing sales in all schools. This letter contained a humorous anecdote on patriotism, 'Mackinder and Herbertson think that 6 or 8 globes representing the whole earth, at the cost of five pounds, would have an enormous market in all English schools. At the end [having heard your name], a journalist wished anxiously to know (patriotism, nationalism) from what area had your father come, whether he was English, Welsh or Scottish'.[142]

Significantly, these sessions were attended by a certain number of persons who could hardly be described as 'geographers', as was the case with the artist Thomas Cobden-Sanderson, who congratulated Reclus, claiming that his contribution went well beyond the cartographic objects presented there: 'It is for that contribution to human knowledge and human aspiration that I beg as a layman to thank M. Reclus for his life-long contribution to the great work of humanity'.[143] In his conclusions, the President also acknowledged Reclus's contribution to the RGS debates, connecting it with British maritime traditions: 'He has reminded us that in the days of the Renaissance [globes] were universally used. Sir Francis Drake and Cavendish and Frobisher certainly took globes like that to sea with them, and puzzled out their problems on the globes, and it is far easier to understand problems in geography by using a globe, than by drawing imaginary spherical triangles on a plane surface'.[144] Imperial and anti-imperial traditions could share spaces among geographical debates at the RGS.

At Reclus's death, the August 1905 issue of *Freedom* was dedicated to his biography, written by Kropotkin, which had first appeared in French for the *Temps Nouveaux* and was quickly translated. Kropotkin also had the idea of republishing this text as a *Freedom* brochure, as such brochures were often printed in addition to the journal, writing to Alfred Marsh that: 'The Jewish comrades[145] have ... a beautiful page-portrait of E. Reclus. I wonder whether we want to keep it very cheap by having 1,000 copies made'.[146] Consistent with their shared programme, Kropotkin considered whether the biography and geographical work of his friend could be used as a possible topic for popular education. This proposed educative pamphlet was not undertaken, but short texts from Reclus, on topics such as education, evolution and political abstentionism were regularly published in *Freedom* over the following years.

It is possible to conclude that their personal experiences in London and in the British Isles played a pivotal role in the political and scholarly training of the Reclus brothers. First, they obtained first-hand knowledge of the social question through witnessing the conditions of the industrial proletariat in Britain and of Irish peasants after the Great Famine. Second, they became politically radicalised within the milieus of French exiles in London where, according to Max Nettlau, 'an anarchist opposition grew up against all the ... authoritarian leaders of French proscription'.[147] Among the most radical exiles, Ernest Coeurderoy and Octave Vauthier authored an influential book, *La Barrière du Combat*. In his recollections to Nettlau, Élie Reclus confirmed that 'he had a clear memory of Coeurderoy and the Vauthier family' in London.[148] Third, integrating their German culture with notions and problems characterising debates within British natural sciences, at the time of the early Darwinian revolution, allowed the Reclus brothers to form a solid foundation for elaborating their theory of mutual aid, together with Kropotkin, and to establish long-life collaborations with British progressive and liberal scholars.

Notes

1   Thompson, *The Making*.
2   Hulse, *Revolutionists in London*, 2.
3   Avrich, *The Russian Anarchists*.
4   Thompson, *The Formation*, 36.
5   Gwynn, *The Huguenots*.
6   Reclus et al. 'On spherical maps and reliefs: A discussion'.
7   Brun, 'Élisée Reclus'.
8   Reclus, *Correspondance, vol. I*, 13.
9   Nettlau, *Eliseo Reclus, vol. I*, 43.
10   Ferretti, 'Comment Élisée Reclus'.
11   Dunbar, *Élisée Reclus*.
12   Reclus, *Correspondance, vol. I*, 51.
13   Reclus, *Correspondance, vol. I*, 51.
14   Reclus, *Correspondance, vol. I*, 52.
15   Bantman, *The French Anarchists*.
16   Di Paola, *The Knights Errant*.
17   Reclus, *Correspondance, vol. I*, 45.
18   Reclus, *Correspondance, vol. I*, 45.
19   Reclus, *Correspondance, vol. I*, 44.
20   Brun, 'Élisée Reclus', 27.
21   Reclus *Correspondance, vol. I*, 47.
22   Reclus *Correspondance, vol. I*, 48.
23   Nettlau, *Eliseo Reclus, vol. I*, 96.
24   Nettlau, *Eliseo Reclus, vol. I*, 97.
25   Nettlau, *Eliseo Reclus, vol. I*, 99.
26   Heath, 'Élisée Reclus', 130–131.
27   Gosudarstvennyi Arkhiv Rossiiskoi Federatsii, Fondy P-1129 (hereafter GARF, 1129), 2, 2692, Heath to Kropotkin, 16 August 1905.
28   Dunbar, '*The History of Geography*', 204.
29   Dunbar, '*The History of Geography*', 205.
30   Dunbar, '*The History of Geography*', 206.
31   Reclus, *NGU, vol. IV*, 673.
32   This chapter's topic is further developed in: Ferretti, 'Political geographies'.
33   Dunbar, '*The History of Geography*', 205.
34   Reclus, *Correspondance, vol. I*, 58.
35   GARF, 1129, 2, 2692, Heath to Kropotkin, 15 July 1905.
36   Ferretti, 'They have the right'.
37   Reclus, *Correspondance, vol. I*, 60.
38   Reclus, *Correspondance, vol. I*, 66.
39   Reclus, *Correspondance, vol. I*, 66.
40   Mächler-Tobar, *Un nombre expoliado*.
41   Reclus, *L'Homme et la Terre* (hereafter *HT*), *vol. III*, 386.
42   Reclus, *HT, vol. III*, 594.
43   Reclus, *HT, vol. IV*, 362.
44   Said, 'Afterword', 177.

45 Dunbar, *The History of Geography*, 12–22.
46 Nettlau, *Eliseo Reclus, vol. I*, 64.
47 This section's discussion of Élie Reclus's anthropological work is further developed in: Ferretti, 'The murderous civilisation'.
48 Dunbar, *The History of Geography*, 122–131.
49 Reclus, *Correspondance, vol. I*, 51.
50 Pierrefitte-sur-Seine, Centre d'Accueil et de Recherche des Archives Nationales, Institut Français d'Histoire Sociale (IFHS), 14 AS 232, Élie and Élisée Reclus to Nefftzer, 6 January 1858.
51 IFHS, 14 AS 232, Élie and Élisée Reclus to Nefftzer, 6 January 1858.
52 Secord, *Visions*, 139.
53 Secord, *Visions*, 156.
54 Secord, *Visions*, 157.
55 Secord, *Visions*, 172.
56 Ferretti, *Lettres*.
57 Livingstone, 'Debating Darwin'.
58 Stocking, *Victorian Anthropology*.
59 Fabietti, *Alle origini*, 36–37.
60 Fabietti, *Alle origini*, 36.
61 Fabietti, *Alle origini*, 17.
62 Fabietti, *Alle origini*, 18.
63 Owen, *Darwin's Apprentice*.
64 Tang, *The Geographic Imagination*.
65 Dugatkin, *The Prince of Evolution*.
66 Ferretti, 'Anarchist geographers and feminism'.
67 Reclus, 'Female kinship'.
68 British Library, Ms Add. 49644 f. 77, E. Reclus to Lubbock, 1875.
69 Stocking, *Victorian Anthropology*.
70 Élisée Reclus, 'Élie Reclus', 241.
71 GARF, 1129, 2, 1308, Scott Keltie to Kropotkin, 5 April 1879.
72 Brun, 'Élisée Reclus', 88.
73 Anthropological Institute, 'President's Address'.
74 Reclus, 'The evil eye'.
75 GARF, 1129, 2, 2293, letters from Reginald Smith to Kropotkin, 1899–1913.
76 Reclus, 'Ethnography and ethnology'.
77 Reclus, 'The fire'.
78 Stocking, 'What's in a name?'
79 Reclus, *Les primitifs*.
80 Reclus, 'Ethnography and ethnology', 613–614.
81 Ferretti, 'Teaching anarchist geographies'.
82 Ferretti, *Élisée Reclus*.
83 Reclus, *Guide du voyageur*.
84 Reclus, *Londres illustré*.
85 Private Collection, Clarisse Brian, *Histoire de Magali*, 60 (2ᵉ année, 9ᵉ'mois, 12 février–12 mars 1862). Thanks to Christophe Brun and Philippe Malburet for giving me a copy of this document.
86 Brun, 'Élisée Reclus'.
87 Reclus, *Correspondance, vol. I*, 224.
88 Reclus, *Correspondance, vol. I*, 226.
89 Reclus, *Correspondance, vol. I*, 231.
90 Kropotkin, *Modern Science and Anarchism*.
91 Graham, *We Do Not Fear Anarchy*.
92 Brun, 'Élisée Reclus'.
93 Reclus, *Correspondance, vol. I*, 332.

94   Reclus, *Correspondance, vol. I*, 333.
95   Reclus, *Correspondance, vol. I*, 334.
96   Ferretti, 'Anarchist geographers and feminism'.
97   Reclus, *The Earth*.
98   Tombs, *The Paris Commune*.
99   Ferretti, *Élisée Reclus*.
100 Dunbar, *Élisée Reclus*, 67.
101 University of Cambridge, Darwin Correspondence Project, Darwin to Hartog Hejs, 21 February 1871, http://www.darwinproject.ac.uk/letter/?docId=letters/DCP-LETT-7500.xml;query=reclus;brand=default
102 *Pall Mall Gazette*, 26 December 1871, 7. Darwin's name as a signatory of this petition is likewise mentioned by: *The Flag of Ireland*, 30 December 1871, 4; *The Morning Advertiser*, 27 December 1871, 5; *The Berkshire Chronicle*, 3 December 1871, 2; *Cambridge Independent Press*, 30 December, 3; *The Glasgow Herald*, 6 January 1972, 2; *The Yorkshire Post and Leeds Intelligencer*, 5 January 1872, 4; *The Greenock Telegraph and Clyde Shipping Gazette*, 2 January 1872, 3; *Newcastle Courant*, 5 January 1872, 6; *Illustrated Times*, 6 January 1872, 2.
103 *Pall Mall Gazette*, 26 December 1871, 7. The list of signatories included the following: 'Lords Amberley and Hobart, Sir John Lubbock, Sir John Rose, Mr. Charles Darwin, Mr. Thomas Hughes, M.P., Professor Fawcett, M.P., Mr. W. D. Christie, Professors. F. D. Maurice, Brewer, &c. Additional adhesions may for the next few days be addressed to Mr. J. M. Ludlow, M.A., 3, Old-square, Lincoln's-inn; or to Mr. Eugene Oswald, 39, Gloucester-crescent, Regent's Park, N.W., 24.'
104 Brun, 'Élisée Reclus'.
105 Reclus, *Correspondance, vol. II*, 93.
106 Ferretti, 'The correspondence'; Nettlau, *Eliseo Reclus, vol. II*.
107 Reclus, *NGU, vol. IV*, 941.
108 Neuchâtel, Bibliothèque Publique et Universitaire (hereafter BPUN), 1990/10, Reclus to Schiffer, 28 August 1878.
109 Reclus, *NGU, vol. IV*, 941.
110 BPUN, 1990/10, Reclus to Schiffer, 28 November [1884].
111 Kearns, 'The political pivot'; Kearns, *Geopolitics*.
112 Ferretti, *Élisée Reclus*.
113 *London Evening Standard*, 29 May 1894, 3.
114 Reclus, *Correspondance, vol. II*, 180.
115 Wallace, *My Life*, 208.
116 Girón, 'Kropotkin between Lamarck and Darwin'; Gould, 'Kropotkin was no crackpot'.
117 Amsterdam, International Institute of Social History (hereafter IISH), Marsh Papers, 22, Kropotkin to Marsh, 22 July 1895.
118 IISH, Marsh Papers, 23, Kropotkin to Marsh, 27 July 1895.
119 Reclus, *Correspondance, vol. II*, 189.
120 Ferretti, 'Geographies of internationalism'; Hodder, Legg and Heffernan, 'Introduction'.
121 Reclus, 'Projet de construction'.
122 Reclus, *Correspondance, vol. II*, 190–191.
123 Ferretti, 'Situated knowledge'.
124 Pearson and Heffernan, 'Million scale mapping'.
125 Reclus, *Correspondance, vol. III*, 100–101.
126 Nettlau, *Eliseo Reclus, vol. II*, 243.
127 'Report of the Holborn Town Hall Meeting'. *Freedom*, August–September 1886, 97.
128 'Evolution, Revolution and the Anarchist Ideal by Élisée Reclus'. *Freedom*, February 1898, 1.

129   Alavoine-Muller, 'Un globe terrestre'.
130   GARF, Fondy 1129, 2, 2103, Reclus to Kropotkin, 13 January 1898.
131   Geddes, Jordan and Brin, 'A great globe', 407.
132   Geddes, Jordan and Brin, 'A great globe', 409.
133   Ferretti, 'Globes'.
134   Wallace, 'The proposed gigantic model'.
135   Wallace, *Studies*, 64.
136   Dunbar, *Élisée Reclus*, 107.
137   Reclus, 'A great globe', 402.
138   Reclus, 'A great globe', 403.
139   Reclus, 'On spherical maps and reliefs', 290.
140   Reclus et al., 'On spherical maps and reliefs: A discussion', 296.
141   Reclus, *Correspondance, vol. III*, 253.
142   Reclus, *Correspondance, vol. III*, 254.
143   Reclus et al., 'A discussion', 297.
144   Reclus et al., 'A discussion', 298–299.
145   An anarchist group using the Yiddish language and which published the journal *Arbeter Freynd* (The Worker's Friend) also printed a Yiddish version of Reclus's pamphlet 'Evolution, Revolution and the Anarchist Ideal'.
146   IISH, Alfred Marsh Collection, 112, Kropotkin to Marsh, 20 July 1905.
147   Nettlau, *Eliseo Reclus, vol. I*, 99.
148   Nettlau, *Eliseo Reclus, vol. I*, 99.

2 Editorial networks and the publics of science

Building pluralist geographies

In this chapter, I analyse the relations between Élisée Reclus, Kropotkin, and their British editors and publishers, undertaking a forensic investigation of unpublished correspondence held in British, French and Russian archives to extend recent works on Kropotkin's relations with publishers such as James Knowles and John Scott Keltie.[1] These works clarify the existence of a conscious strategy deployed by Kropotkin to use print cultures to 'sell anarchism to the British'[2] and to work as a virtual full-time anarchist propagandist thanks to his paid 'scientific work', which is not strictly distinguishable from Kropotkin's 'political' outputs. Kropotkin employed a strategy likewise deployed in France by Reclus with the publishers Hetzel and Hachette.[3] Reconstructing British editorial networks of the anarchist geographers is fundamental to understanding not only the importance of places, contexts and circulations for the construction of knowledge, including political ideas, but also contributes to the existing scholarship highlighting the role played by publishing businesses and the book industry generally for shaping early geographies.

The contentions advanced in these chapters are closely connected with the social and technological innovations which characterised the print industry at a time which saw a steady increase in sales, profits and readerships concerning both scientific books and the periodical press throughout the nineteenth century and beyond.[4] These developments also corresponded to innovations in the 'materialist hermeneutics' of the printed page, which was rendered increasingly readable and accessible for all types of readers, including those reading geographers' works,[5] and paralleled international trends, as in the French case, where the extraordinary success of Reclus's *NGU* was possible thanks to the spectacular democratisation involved in cultivating the widest possible readership which large publishing industries such as Hachette were promoting. Democratisation of the reading public and increasing the numbers of readers proved an attractive approach for both commercial and political reasons. These wider social processes allowed the anarchist geographers to earn their living without the need of university chairs, and to disseminate their political views to wider audiences beyond activist circles.

In this chapter, my main argument has been that scholarly, activist and editorial activities were closely interconnected in Kropotkin's and Élisée

Reclus's experiences in Britain, and that participation in the business of editorial activity was one of the principal reasons for their success within British scholarly milieus. While scholars such as Gerry Kearns have rightly highlighted the value of scientific tolerance exhibited by RGS members in welcoming refugees and unorthodox authors such as Reclus and Kropotkin, I show below that this was due not only to the disinterested open-mindedness of these British scholars, but also to the fact that they were almost all involved in editorial business activities where Kropotkin's and Reclus's work was considered as a resource for the scientific press and for endeavours such as the *Encyclopaedia Britannica*. Thus, the idea of a 'bargain' as mentioned earlier, finds support in the promotion of their interests that the participants in these exchanges could bring about simultaneously through mutually beneficial editorial, scholarly and political collaborations.

Kropotkin's early networks: John Scott Keltie, Patrick Geddes and Joseph Cowen

Arriving in Britain in 1876 after his escape from prison in Saint-Petersburg in Russia, Kropotkin was first welcomed and supported by John Scott Keltie (1840–1927), who remained a collaborator and a life-long friend of Kropotkin. The story of their first meeting is told by Kropotkin in his *Memoirs* and their professional relationship has already been the object of works carried out through investigating relevant archives and correspondence.[6] Therefore, I do not focus on what has already been stated concerning relations between Kropotkin and Keltie, but rather focus on some aspects which have been less examined, such as Keltie's involvement with the scholarly and activist networks which supported Kropotkin after his arrest in France in 1882. This involvement was part of the 'Scottish connection' of Kropotkin and the Recluses, which I address in the last chapter and which included the Edinburgh networks of Anne and Patrick Geddes, who were scholars and social reformers who collaborated with the anarchist geographers for 50 years, from their first contacts with Kropotkin in 1882 to Geddes's death in France in 1932, while he was still working with Paul Reclus for the construction of a geographical museum in south-west France, modelled on Edinburgh's Outlook Tower.[7]

Geddes's archives in the National Library of Scotland show that Kropotkin lectured in Edinburgh during 'a tour on Russia'[8] and met the Geddeses prior to being imprisoned in France from December 1882 to January 1886, charged mainly with professing anarchist ideas, a charge to which Reclus publicly declared himself guilty and wrote to the Lyons judge who had convicted Kropotkin requesting (unsuccessfully) to be jailed like his friend.[9] The correspondence between Scott Keltie and Geddes reveals the early and critical role that both scholars played in collecting signatures for another petition sent by British scholars to the French government, this time in favour of Kropotkin. In late December 1882, Keltie wrote to Geddes: 'Poor Kropotkin: what can we do? I have just seen a Russian friend of his[10] and there is no doubt the whole thing is set up ... If I can do anything to get him out of jail. Robertson Smith might be able to do something'.[11] What is significant here is that important elements of the scholarly British world seemed to be able to mobilise quickly in favour of Kropotkin without any doubt or any questioning. Those mobilised did not appear to believe in the allegations of the French justice system against Kropotkin, and did not express doubts that something needed to be done for him. In February 1883, Keltie asked Geddes for news on the results of the 'Kropotkin petition'[12] in Edinburgh. Some weeks later, Keltie sent to Geddes 'the complementary sheets of the Kropotkin petition, containing ... some good names', requesting him to sign it and to 'get Mackay do so and such of the *Britannica* men',[13] which included Kropotkin's acquaintances such as John Black and William Robertson-Smith. The back of the letter had been filled in, by either Geddes himself or one of his collaborators, with a list of Scottish signatories which included major figures of scholarship and publishing such as John Murray, Robert Chambers, Archibald and James Geikie and several university lecturers and professors, while Keltie provided a list of English supporters which included Colonel Yule, Henry Woodward, Henry Bates, William Morris, Alfred Wallace, and many others.

Socialist leader Henry Hyndman (1842–1921), likewise Kropotkin's friend and correspondent, provided a tenor of this petition in his recollections:

> I went into Earl's Court Station ... with Tchaikovsky the Russian Anarchist. We had gone to [Colonel] Yule in order to obtain his signature to a Memorial to the French Government I was using my best endeavours to get signed in favour of giving Kropotkin, then imprisoned at Clairvaux for complicity in Anarchist plots, better accommodation, and the right to see his wife. The names appended to that Memorial were those of some of the most distinguished men of science and men of letters in the country, which I think did them great credit ... The man who refused most positively to lend any help at all was, as it happened, Thomas Huxley, who gave it as his opinion that Kropotkin was already too well off as he was. ... I had previously made Prince Kropotkin's acquaintance and friendship upon an introduction from Joseph Cowen, that consistent friend of the revolutionists and subversionists in every country.[14]

Thomas Huxley (1825–1895) was a significant exception in the otherwise supportive British scientific world, and it is no coincidence that his 'social Darwinism' would be one of Kropotkin's targets when he expounded to the British public his theories on mutual aid as a factor in evolution in the following decades.

Evidence from Kropotkin's archives show that, during his imprisonment in France from December 1882 to January 1886, the Russian geographer continued to work ceaselessly for Reclus's geographical projects and for British journals, and that he remained in particularly friendly relations with his former British correspondents, especially Keltie, then sub-editor of *Nature* and principal employer of Kropotkin.[15] During these same years, some of the correspondence between Kropotkin and *Nature* was dealt with by another editor and student of folk-lore, A. Granger Hutt on behalf of Scott Keltie. The notably friendly tone of several of these letters, sent in 1883 to Kropotkin's jail, further indicates the support he received from the British editorial and intellectual world despite being persecuted no longer by an autocratic 'Eastern' state, but by a Western democracy allied with Great Britain.[16] Therefore, just as they had done 11 years beforehand with Reclus, British scholars and activist militants networked to gather support for Kropotkin. In both cases, pleading the cause of political prisoners seems to have been justified not exclusively as an expression of a liberal mentality and the presence of a certain number of scholars sympathising with progressive political views, but also by a degree of British 'isolationism' that deemed British political liberties as always better than those available on the continent, as also Reclus noticed in his works.[17] Defending a subversive jailed in Russia or France was doubtlessly less compromising than doing the same if that person were to have been imprisoned in England.

As noted above, Kropotkin was introduced to Henry Hyndman by Joseph Cowen (1829–1900). Cowen was a distinctive figure as a radical politician, MP for Newcastle until the 1886 Home Rule crisis and the fall of the government of William Ewart Gladstone (1809–1898),[18] but most importantly the editor of the *Newcastle Daily Chronicle*. Dealing with the daily press was not an unusual exercise for scholars such as Reclus and Kropotkin. Correspondence between Cowen and Kropotkin started in the summer of 1881, apparently after Kropotkin contacted Cowen to volunteer giving lectures on Russia's political situation within Cowen's political circuits 'to put his case before an English audience'.[19] Cowen responded, accepting the idea, but delaying its realisation because he considered that organising lectures in the summer was not suitable: 'I do not think that lectures at this season of the year could be convenient or advantageously achieved … If it could be arranged that you could deliver a series of lectures in November, I think it wold be of grand service, and you might rely upon me doing everything in my power to assist you'.[20] While the projected cycle of lectures was apparently not realised that year, in September Cowen wrote to Kropotkin to say that he would be 'glad to receive at your convenience the … article on Russia you proposed to write for the *Newcastle Chronicle*'.[21] At that time, their personal acquaintance in London is shown by invitations for dinner sent by Cowen.[22]

Kropotkin published a series of 19 short articles for the *Newcastle Daily Chronicle* from 12 October 1881 to 7 February 1882, all related to Russian social and political issues such as the situation of political dissidents, the famine and the thought of Leo Tolstoy.[23] According to Martin Miller, these articles played a vital role in introducing Kropotkin to British popular readers, among whom his denunciation of political and social conditions in Russia became rather popular. In his letters, Cowen expressed the opinion of most of

Kropotkin's British friends about Kropotkin's frequent stays in Switzerland undertaken to propagate his political views, and his project of settling in Thonon, a small French town on Lake Geneva, following his expulsion from Switzerland in 1881: 'I am pleased to hear that you have got safe to London and I trust that you will find residence there a little pleasanter than in the small French town'.[24] However, according to most of his biographers, Kropotkin was unhappy with his life in England in those years, and chose to move back to the continent in autumn 1882, despite the risks of political repression, and the contrary advice of Keltie, Cowen and others. In 1882, Kropotkin considered that: 'Better a French prison than this grave'.[25] By 1886, as I explain below, his views had become very different.

As in the example of Keltie's case, personal relations and trust were fundamental in building Kropotkin's professional relations. Eventually, Cowen proved his friendship by seeking paid work for Kropotkin in American newspapers such as the *New York Herald*, but at the same time he warned the Russian prince: 'It may be desirable for you to bear in mind that ... newspaper people generally have no sympathy with your cause, and they simply seek to utilise you for purposes of notoriety or gain'.[26] Indeed, Cowen was one of the few British editors collaborating with Kropotkin who also had some sympathies for his ideas. In 1882, he forwarded to other British MPs a pamphlet promoted by Kropotkin and Tchaikovsky against Russian autocracy.[27] In July of the same year, he invited Kropotkin to a meeting in Newcastle attended by local miners. With some perhaps involuntary sarcasm, Cowen wrote to the prince that there, he 'would get a better idea of the working people'.[28] However, a common theme of Cowen's letters to Kropotkin were the apologies that the MP sent justifying himself for not attending Kropotkin's lectures due to his parliamentary commitments, but asserting in compensation: 'I sincerely sympathise with you in your efforts to spread information as to the condition of your countrymen. You will find no feeling anywhere, in England, against the Russian people ... Our enemy is despotism'.[29] Again, British solidarity for Kropotkin and Russia can be perceived as enlightened, but also as nationalistic.

After expressing his sadness at Kropotkin's decision to leave for France, Cowen responded quickly when the news spread that Kropotkin had been arrested, writing to Kropotkin's wife Sofia to offer a 'few words of sympathy for you both in your great trouble and an inquiry if I could be of any service to you or your husband ... What can be done to aid your husband in his trial and to get him out of prison?'.[30] Furthermore, he sent his collaborator MacDonald to Geneva 'to see you [Sofia] and your husband and learn whether there is anything we can do'.[31] Cowen promoted a petition to aid Kropotkin in collaboration with Tchaikovsky, and his offers of assistance to the prisoner included financial help, and apparently a bail amount which was refused by Kropotkin.[32] Cowen wrote then to a colleague that: 'We have been getting up a memorial from English literary scientific men, to the French government, in favour of Kropotkin, in the same way as a memorial was got up in favour of

M. Reclus. We have got it signed by some of the best men in this country ... and it will be ready to be dispatched to Paris in the course of next week.'[33] As with the case of Reclus, scientific diplomacy needed to find sympathetic persons in the political world, which proved again to be difficult. While Cowen conjectured that French radical MP Georges Clemenceau (1841–1929) would have been interested in the case, he complained that he did not have any direct contact with French ministers 'except M. Challemel-Lacour, with whom I have a slight acquaintance'.[34] Paul-Armand Challemel-Lacour (1827–1896), a long-time republican and opponent of the Second Empire, was likewise an acquaintance of the Recluses.[35] Nevertheless, a rift between the anarchists and the left-republicans in France had already become very deep, and the only result of the petition was that Kropotkin was moved from Lyon to Clairvaux, closer to Paris, where he had more possibilities for working and receiving books.[36] Eventually, all his British correspondents expressed solidarity with Kropotkin in arguing for a sort of 'inferiority' of French institutions, more associated with Russian autocracy, as opposed to an allegedly freer England. This is evidenced in Cowen's case, in which he blamed the 'savage sentence' and the 'cruel imprisonment'[37] the Russian prince had received, and also volunteering to visit Kropotkin in his prison, if this 'could help you in any way'.[38]

On Kropotkin's return to England in 1886, correspondence with Cowen restarted, with mention made of a payment Cowen had to make to Kropotkin, but with no specific paper identified, which suggests the likelihood of new projected lectures, for which 'a decent remuneration'[39] was expected. Kropotkin continued to see Cowen in Newcastle, stopping there during his trips to Edinburgh,[40] and always remained on friendly terms with him, although with no further formal collaboration. From 1886, a time of social and scholarly respectability had arrived for Kropotkin.

Geographers, editors and gentlemen: Henry Bates, Hugh Mill, William Robertson-Smith, Hugh Chisholm

After his work for *Nature*, the next most important contribution to British scholarship Kropotkin made was to the ninth edition of the *Encyclopaedia Britannica*. When Keltie put Kropotkin in touch with William Robertson Smith (1846–1894), then editor of the *Encyclopaedia*, with which mutual friends such as Augustus Henry Keane (1833–1912)[41] were already involved, the sub-editor of *Nature* wrote to the Russian prince that Robertson Smith was 'a Scotch clergyman ... a professor in the Church there but was recently ejected for heresy', a remark which was likely to have stimulated the curiosity of the anarchist geographer.[42] According to Livingstone, Robertson Smith's works on evolution were another case of the adaptation of Darwin's theories to specific philosophical and political stances. In this case, the attempt had involved trying to find 'revelation in anthropology'; that is, reconciling evolution and Christianity.[43] Between 1883 and 1888, Kropotkin wrote twenty-eight articles for the ninth edition of the *Encyclopaedia Britannica*. [44] Unfortunately, the surviving letters Robertson Smith sent to him did not directly address evolution, but they do show an unexpectedly close relationship between the anarchist and a clergyman. Indeed, Robertson Smith introduced Kropotkin to Archibald Geikie (1835–1924) considered at that time, together with his brother James (1839–1915), as among the champions of free thought and supporters of science against religion, positions that the anarchist prince clearly endorsed. [45]

Furthermore, Robertson-Smith's letters to Kropotkin clarify a point in Kropotkin's biography concerning a university chair that the Russian geographer had allegedly refused to accept, to remain consistent with his anarchist ideals. This story was first recounted in James Mavor's recollections and has been critically discussed by Woodcock and Avakumović. According to Mavor, Robertson Smith 'was on very intimate terms with Prince Kropotkin, and ... was anxious to secure Kropotkin for Cambridge as Professor of Geography. Kropotkin told me that he did not care to compromise his freedom by accepting such a position; but he felt very pleased that Robertson Smith's friendship had prompted him to so generous a project'. [46] The letters sent by the Scottish scholar do not contain any offer of this kind, but only mention some suggestions about an interest that Kropotkin had expressed in lecturing for an extension programme in Cambridge. The more significant advice that Robertson Smith gave to Kropotkin when they met in Cambridge was to avoid 'political and social lectures', even though he knew that this went against the Russian's activist inclinations. [47] This understanding is confirmed indirectly by Keltie's letters to Kropotkin, when, some years later, Kropotkin expressed interest in lecturing for an extension scheme at Oxford, and Keltie responded that: 'I have reasons to think that if you care about your ... programme you could probably be accepted. But ... you must cut all socialistic lectures'. [48] Recent literature has likewise shown that anarchist geographers such as Reclus and Metchnikoff were far from being opposed to university teaching on reasons of principle. [49]

Kropotkin corresponded as well with the assistant editor of the *Encyclopaedia Britannica*, John Sutherland Black (1846–1923), who was also a Scot and a biblical scholar. Their exchanges started in the summer of 1882, when Kropotkin was still in Britain, [50] and continued with a certain regularity and with always the same courtesy during Kropotkin's captivity in France. Therefore, it is possible to conclude that, from 1883 to 1885, Kropotkin's cell in Clairvaux was a place where editing work for parts of the main European encyclopaedical works of that time, such as Reclus's *NGU* and the *Encyclopaedia Britannica,* was being undertaken. From the tone of communications, one also has the impression that this location was considered as a workplace like any other, regardless its material constraints, evidenced with a note from Black urging the prisoner to send him the entry on the Russian town of Perm: 'My dear Sir, the printers are now in urgent need of the manuscript for Perm (province and town)'. [51] Indeed, his situation did not constrain Kropotkin's working potential. In addition to some dozen entries on major cities and regions in Russia, Kropotkin proposed to the editors of the *Encyclopaedia Britannica* some further material on small towns, which was politely rejected by Black, who stated that the additions offered were 'not likely to be consulted by many of our readers'.[52] In any case, the frequency of these exchanges prior to Kropotkin's return to England in 1886 shows that the work done for the *Encyclopaedia Britannica* was demanding, and also remunerative for the exiled scholar until 1887.

Kropotkin's collaboration was additionally requested for the tenth edition of the *Encyclopaedia Britannica*, again due to Keltie, as the Scottish geographer had been appointed sub-editor for the section on geography and statistics. Keltie asked Kropotkin to write 'the articles on Russia ... The main work would be to bring the articles up to date from the time they were written'.[53] Kropotkin was reluctant to do this work, primarily because he had no interest in what he deemed to be essentially a compiling exercise, but he finally accepted and was paid at the rate of '£4 per page'.[54] Some difficulties developed when the extent of Kropotkin's work exceeded the number of pages Keltie was permitted to use, as set by the principal editor, Wallace Mackenzie. This entailed editorial intervention, although not in terms of political control. For example, Keltie asked Kropotkin, with some humour, if he deemed it necessary 'to mention in detail all the different kinds of shells of the Aral and Caspian seas'.[55] Undeterred, and clearly appreciated, Kropotkin's work for the following editions of the *Britannica* consisted not only in updating existing entries, but also in producing new material.

The most famous and politically significant of Kropotkin's contributions for the *Encyclopaedia Britannica* occurred with the eleventh edition, edited by Hugh Chisholm. Agreed in 1905, his contributions consisted primarily in an updating of most of the entries on Russia which Kropotkin had written for the previous editions, while approximately 'one-third would require re-writing'.[56] Nonetheless, the most important novelty of this edition was the entry on 'anarchy' completed in 1910. This text was widely circulated as a brochure of anarchist propaganda independently from its original editorial purpose, as it contained a complete explanation of the history and theoretical tendencies of anarchism until the present, summarized in only 5,000 words. Clearly, the editor had insisted on brevity, as shown by a letter from Kropotkin claiming that he could 'not see any way to make it shorter'.[57] However, the most unusual aspect of the editorial correspondence between Kropotkin and Chisholm involved a letter in which Chisholm announced his insertion of a note which aimed to reassure the British public concerning anarchy, commonly considered as an evil in bourgeois milieus. 'I have inserted a rather long footnote – for which you will bear no responsibility – giving, for historical purposes (as this is the heading under which it must come) a rather more detailed account than your general reference as to the more startling "anarchist" outrages of the last 25 years. I have prefaced it by reminding people that "anarchist" is a term used very loosely by the public'.[58] It is worth noting that Kropotkin, like Reclus,[59] was totally opposed to so-called anarchist individualists responsible for dynamite attacks in France in the 1890s.[60] The fight against 'Ravacholism' (after French individualist Ravachol, the term indicating the practice of indiscriminate violence), and against the spread of amoral individualistic tendencies inspired by authors such as Stirner and Nietzsche among anarchists, informed a generation of international activists, especially Errico Malatesta and Luigi Fabbri.[61]

Trying to avoid at the same time 'compromising' the *Encyclopaedia* with anarchism and disappointing the anarchist prince, Chisholm's diplomatic efforts were admirable, and no protestation from Kropotkin followed in the surviving correspondence. Ironically, the point on which Kropotkin claimed that 'the last decision must be left with me'[62] was not about political contents, but about the appointment of a consultant on geological matters proposed by Chisholm, with the task of peer-reviewing Kropotkin's works on Russian physical geography. Nevertheless, Kropotkin concluded the discussion writing that, on editorial matters, he 'fully trust[ed Chisholm's] eye'.[63] This short series of exchanges exemplifies perfectly the nature of the 'bargains' established between anarchist geographers and their liberal editorial and scholarly colleagues and employers, as explained in the introduction.

These editorial endeavours were often associated with the activities of scholarly bodies and learned societies. One of the most important and well known of Kropotkin's connections with British publishing and scholarly milieus involved the RGS, through his friend Keltie but also thanks to Keltie's predecessor as RGS secretary, Henry Walter Bates (1825–1892). Naturalist, evolutionary thinker and a comrade of Wallace in his 1848–1852 Amazon expedition,[64] Bates worked for the RGS as an assistant secretary from 1864 and, according to Kropotkin's recollections, he was an important inspiration for the anarchist prince's approach to evolution. Two persons encouraged Kropotkin to write on mutual mid, the first being James Knowles, while

> The other was H.W. Bates, whom Darwin has truly described in his autobiography as one of the most intelligent men whom he ever met. He was secretary of the Geographical Society, and I knew him. When I spoke to him of my intention he was delighted with it. 'Yes, most assuredly write it', he said. 'That is true Darwinism. It is a shame to think of what "they" have made of Darwin's ideas. Write it, and when you have published it, I will write you a letter in that sense which you may publish'. I could not have had better encouragement, and began the work which was published in *The Nineteenth Century* under the titles of 'Mutual Aid'.[65]

Thus, Bates became one of Kropotkin's supporters in his campaign against Huxley's and the social Darwinists' interpretations of evolution, together with the most progressive elements of the RGS, as discussed by Kearns, who claims that: 'Bates, Keltie, Freshfield and Mill, central to the work of the RGS, shared Kropotkin's aversion to the social Darwinism of Huxley, and indeed of Mackinder'.[66]

In a biographical study of Bates, John Dickenson highlights the important roles he played in fostering RGS activities, in its collaboration with colleagues such as Keltie and Mackinder, and in editing books on exploration and text-books for educational programmes: 'Clements Markham ... was a prime advocate of the explorer tradition of the RGS. However, there were other members who sought to introduce a new dimension to its activities, in fostering geographical education and scientific geography. Early efforts saw the award of prizes for examination in geography at schools, over the period 1869–84; Bates served as an examiner.'[67] Bates had entered into correspondence with Kropotkin in November 1882, before Kropotkin's imprisonment in France, thanks to Keltie: 'Our mutual friend M. Keltie has referred to you as a gentleman ... able to furnish the Society with information regarding Mr. [Pavel Mikhailovič] Lessar's recent journey from Kabul to Merv, an account of which has appeared in the *Golos*. ... We pay for such contributions, *bien entendu*.'[68] Although it was well known that Kropotkin was more comfortable with French than with English and that during many years his papers in English needed serious copyediting by native speakers, Bates was not ungenerous in 'complimenting [Kropotkin] on the mastery [he] show[ed] on our language'[69] after receiving that text.

This epistolary exchange inaugurated a 30-year period of contributions from Kropotkin for publications of the RGS such as the *Proceedings* and the *Geographical Journal*, later edited by Keltie, Mill and other correspondents of Kropotkin. Kropotkin's multilingualism and field experience opened for British geographers an immense opportunity for indirect exploration and the possibility of reading the accounts of Russian, German and French travellers in Siberia and Central Asia, with a clear imperial interest for RGS members.[70] For Kropotkin, while the primary interest in producing these notes and translations was doubtlessly material, his work can also be considered as an indirect means for making the Russian situation more familiar to the Western public. Relations between geography and politics were, however, ambiguous, as a formal rejection of politicisation was contradicted by its explicit presence during discussions on explorations at the RGS, as Bates wrote to Kropotkin about the presentation of Kropotkin's paper on Lessar's journey: 'It was read on Monday November 27th and excited a brilliant discussion, I am afraid more political than geographical'.[71] Bates' correspondence apparently stopped during Kropotkin's prison years, but restarted in 1888.

Woodcock and Avakumović claim that Bates volunteered to write a preface for the book edition of *Mutual Aid*, 'endorsing its main contents and its attack on Huxley's distortion of Darwinism. But he was dead before the first edition was ready.'[72] This case confirms that, for Kropotkin, nothing was neutral in science, and that an otherwise not overtly politicised scientist like Bates was prepared to participate in Kropotkin's strategy to foster his ideas on evolution.

Librarian at the RGS from 1885, Keltie became secretary in 1892, succeeding Bates, and the librarian's duties were then entrusted to Hugh Robert

Mill (1861–1950), who had occasion to correspond with both Reclus and Kropotkin in this capacity. Scottish, trained in Edinburgh, oceanographer and specialising in meteorology, well ensconced in British editorial milieus, Mill served as RGS librarian from 1892 to 1900.[73] As with Kropotkin, he was also introduced to the university extension programme by Patrick Geddes, and published an important volume on physiography endorsed by Kropotkin[74] and at least two useful texts which help understand many working aspects of the RGS at that time, namely, an obituary of Scott Keltie[75] and an *Autobiography.*[76] Kropotkin's letters to Mill, surviving at the RGS archives, show that the librarian was also committed to editing Kropotkin's papers for the Society's publications. An 1892 letter reveals how the Russian prince continued to feel uncomfortable with the daily requirements of written English, apologising for 'a mistake which a still imperfect knowledge of English had made an overlook in the Ms and the proof. The sources instead of springs is again one of those mistakes. I am extremely grateful to you for having indicated me those mistakes and will always be the most grateful in the future'.[77] The need for copyediting and editorial support undoubtedly encouraged the Russian geographer to ensure a wide range of collaborative networks, as his need in this regard is mentioned on many occasions.

Mill served both to provide working materials for Kropotkin, who acknowledged for example the shipping of a number of pamphlets in 1893,[78] and as a proof-reader, for example, for the series of papers Kropotkin started to publish for the *Geographical Journal* in that same year.[79] Kropotkin's accounts of the physical phenomena that he had studied in Siberia seemed to particularly interest Mill,[80] who regularly provided Kropotkin with scientific texts to review.[81] However, the most important document to indicate the extent of the collaboration between Mill and Kropotkin in its social and political aspects is a programme of 30 lectures for an extension programme, subdivided into three main topics, which Kropotkin submitted to Mill in November 1893:

> The subjects I propose to submit to the Universities for U[niversity] E[xtension] lectures are: 1. The Ice-Age, chiefly in regard to the proofs of its occurrence ... (10 lectures); 2. The structure of Central and Northern Asia ... (10 lectures); 3. The origin and evolution of institutions from mutual protection and support: in primitive societies; in barbarian ages; in medieval times; in modern times (10 lectures).[82]

This indicates that, after having been strongly discouraged by Robertson-Smith and Keltie, Kropotkin tried again to apply for university extension lecturing, including in this programme the essential elements of his scientific contribution to that time. Although Mill's response to this letter does not survive, it could be surmised that discussions on mutual aid topics would likely have been 'socialist lectures'. The Ice-Age topic would probably have been acknowledged as non-political, but physical geography and pre-history also had a political value for the early anarchist geographers in countering religion and Biblical tales.[83] Regardless of outcome, this sort of syllabus confirms the centrality of popular dissemination of scholarly work for anarchist geographers like Kropotkin.

From 1898, Kropotkin contributed to various editions of Mill's *International Geography*, a gazetteer which was first published in 1899 with the collaboration of 70 authors, including Scott Keltie, Ravenstein, Herbertson, and Keane, comprising a network of liberal thinkers involved in the RGS and in associations for the teaching of geography such as the British Association for the Advancement of Science.[84] In 1898, Mill wrote to Kropotkin that he was unhappy with the contributor appointed for a chapter on 'the Russian Empire'. As Kropotkin had already declined to write this chapter, Mill asked, as a sort of compensation, that the anarchist prince 'recommend me someone'.[85] Kropotkin suggested David Aïtoff (1854–1933), a Russian cartographer living in Paris and working with Hachette, and also in collaboration with Franz Scharder and Onésime Reclus. His work on the chapter gave Mill 'the greatest satisfaction',[86] although Kropotkin was still informally solicited to provide corrections to the chapter for the following editions, especially regarding statistics.[87] The details of this engagement (which did not appear in any formal acknowledgement in the written edition), reflected the networking synergies operating in the same way for geographers working at Hachette's, which can be seen as more similar to an informal application of the principles of mutual aid than to contractual obligations.[88] In 1901, it was Kropotkin's turn to request Mill's help for his *Encyclopaedia Britannica* chapter on Mongolia, by asking for bibliographic insight to resolve doubts about the so-called 'red deposits'.[89] These working methods also help to make sense of the interest which anarchist geographers had in accessing the comprehensive possibilities offered by well-endowed distribution facilities, such as a publisher like Hachette or the RGS, to advance their scholarly discourses.

Mill was also a British contact for Élisée Reclus. In 1895, after his 'divorce' with his main publisher Hachette, Reclus tried to continue his editorial business projects in Britain, seeking a publisher for an updated version of the *NGU* and for his final work *Social Geography* [90] which would be published in France with the title *L'Homme et la Terre* (1905–1908) and has regrettably never been translated into English. This need to find a publisher became an occasion for corresponding with Mill, whom Reclus had met at the 1895 International Geographical Congress for which Mill, together with Keltie, had been a principal organiser, as revealed in letters Mill sent to Kropotkin that year.[91] Mill was additionally committed to helping Reclus concerning copyediting of papers which the French geographer started to publish in English in the 1890s. In April 1894, Reclus wrote to Mill: 'My *Review* article is finished and I think that I can send you a copy tomorrow. You might deem some remarks rather extraneous to geography, but *tout se tient*!'.[92] This letter was in reference to an article 'Russia, Mongolia and China' that Reclus published for the *Contemporary Review* in May 1895. What is significant is that

Mill's technical work of copyediting was the occasion for a discussion between the two men on the disciplinary status of geography.

As Mill's answers do not survive to the best of my knowledge, I can only try to infer Mill's views based on Reclus's letters. Reclus had a very large conception of geography and was uncomfortable with narrow disciplinary boundaries. Apparently, Mill had more traditional views on what could be considered 'geography'. In a letter sent in November 1895, Reclus wrote:

> Anyway, whether it is history, anthropology or geography strictly defined, you would be an excellent translator and you will give me great help as an annotator and collaborator. I start immediately considering your remarks about the need of introducing in maps' captions some summaries on peoples' exoduses and migrations. I completely agree on this. I also thank you for the precious remark on the very secular spirit which resided for such a long time among the Anglo-Saxons and especially among the Scottish.[93]

Reclus's correspondence with Mill also confirms the reputation of the French geographer as a tireless worker, especially in the case of a letter where Reclus revealed his intended work schedule for Christmas 1895: 'I might be free for going to England during Christmas holidays period, from 20 December to 10 January. But will I have any chance to treat some business? Would not everybody be snuggled beside the family fire?'.[94] Undertaking editorial business was of overriding concern. This need is crucial to understanding the anarchist geographers' long-standing involvement with the RGS.

The Nineteenth Century: socialism and evolutionary theorising[95]

James Knowles (1831–1908), a major influence within Victorian cultural scenes as the editor of *The Nineteenth Century,* was undoubtedly the most important of Kropotkin's British periodical publishers. In letters he sent to Kropotkin when the anarchist prince was imprisoned in France, Knowles protested in similar fashion to other British scholars concerning French and Russian authoritarianism and in inviting Kropotkin to return to the homeland of liberalism in Britain.[96] A progressive, a friend of Gladstone and with some interest in socialism,[97] Knowles was also undoubtedly one of Kropotkin's most important employers, having commissioned from him more than 30 lengthy articles starting in 1882.

In some cases, proposals for article topics came from the editor, while in other cases they came from Kropotkin, although Knowles's letters show that in almost all cases the proposals came from the Russian geographer. On being sent an article called 'Petty trades and Grand Factories', Knowles declared that he was 'delighted to receive [it]', as well as a subsequent 'manuscript on the Integration of Labour'.[98] These pieces of work later served as bases for Kropotkin's book *Fields, Factories and Workshops* [99] discussed in the next chapter. This subject matter set the terms for regular discussions between Kropotkin and Knowles, on the copyright of the work and the possibility of re-using the material, or part of it, to produce books. This type of consideration arose because these articles were most often organised as a series, which fostered the idea of reworking the respective topics into book form, as was the case with *Russian and French Prisons, Modern Science and Anarchism, Mutual Aid, The Great French Revolution* and later *Ethics*. Knowles claimed exclusive rights over the commercial exploitation of these texts. There is evidence here of a first clash occurring between the interests of the anarchist propagandist, promoting the widest circulation of his writings, and his editor, more interested in sales and asserting his rights in respect of commercial gain as guaranteed by the copyright laws, warning Kropotkin that 'the copyright of these articles legally belongs to me for 28 years'.[100] Finally, and despite these difficulties, Kropotkin managed to produce all the books he wished, thanks to his editorial success secured through the enormous readership he was gaining through *The Nineteenth Century*, then the most important British monthly by number of printed copies, with around 20 thousand sold for each number in the 1880s.[101]

The most important of Knowles's initiatives in dealing with Kropotkin's output concerned his awareness of the editorial potentialities of Kropotkin's work on mutual aid, whose publication had been refused in principle by another publisher contacted by Keltie. Indeed, in 1881, Keltie had expressed his incomprehension regarding a book that Kropotkin had proposed in which is found an early description of the idea of mutual aid. One of the publishers consulted by Keltie was 'very doubtful about your proposed book on the Law of Mutual Help and can hardly give a definitive answer before seeing your article'.[102] Thus, while Keltie and his contacts had been unable to grasp the editorial potential of the idea of mutual aid, Knowles accepted it enthusiastically, writing in 1889: 'Let me see the other articles on "Mutual animal help". The subject is exceptionally interesting and the treatment at your hands will be equally so'.[103] In *The Nineteenth Century*, Kropotkin's papers on mutual aid offered an alternative to the writings by Knowles's friend Huxley and challenged Huxley's idea of competition by considering cooperation as a part of evolutionary processes. This meant that anarchist geographers like Kropotkin had taken sides in political battles over science between liberal and conservative elements of British society.[104]

The relationship between Knowles and Kropotkin came to an end when a rift developed over the 1905 Russian revolution. While Knowles celebrated the 'great news' that Kropotkin's 'articles may now be circulated in Russia', he restricted their publication in Britain.[105] A 1906 article by Kropotkin denouncing the 'White terror' that followed the 1905 revolution was judged by Knowles as 'terrible and awful. Too terrible for me to publish it as it stands'.[106] Kropotkin's responses, surviving in Knowles's archives in Westminster, show that the Russian prince, although 'awfully disappointed about the article', was disposed not only to turn down his earnings but also to collect money for the hypothetical expenses that Knowles feared in case of legal actions by the Russian embassy.[107] Kropotkin's urgent political agenda involved raising awareness of liberal public opinion to the repression occurring in Russia and he considered the editor's unwillingness to publish as morally unacceptable behaviour.[108] A rupture was inevitable and, in the end, Knowles wrote to Kropotkin that he hoped they could maintain 'personal relations'.[109] However, no more contacts are documented between the two men. Although it appeared that in times of revolution, editorial 'bargains' could not satisfy the needs of Kropotkin's political agenda, his bargain with Knowles had allowed the anarchist prince to be launched editorially as a major force and to advance publications on anarchism for a wide readership in late Victorian and Edwardian Britain.

In 1908, after Knowles's death, Kropotkin wrote to his widow saying that Knowles had been 'one of [his] best personal friends in England', and restarted a collaboration with the editor's son-in-law and successor at *The Nineteenth Century*, William Skilbeck, with whom he published his final series of articles on mutual aid.[110] Contact between Kropotkin and Skilbeck arose on the occasion of a cover letter Kropotkin had written on behalf of a certain Christaleff in support of a proposed paper of the latter for *The Nineteenth Century*. The editor's answer included an explicit request to Kropotkin to restart direct collaboration. 'I confess to a feeling of disappointment when I opened a letter … in your handwriting and found an article by anyone else than yourself. It is too long since you wrote in this review … You have probably not entirely forgotten meeting me occasionally at lunch with my father-in-law'.[111] Recalling Kropotkin's *Recent Science* series[112] in a letter a few days later, Skilbeck asked whether there was 'any hope of your resuming that work'.[113] Kropotkin's answer confirmed his complete reconciliation with *The Nineteenth Century*, as well as providing an opportunity to acknowledge the value of his editorial experience for disseminating his ideas: 'I was so glad to see once more your handwriting which so much reminded me that of our dear Sir Knowles, our dear departed friend [and to] know that you are the Editor of the Review which consequently will continue to retain its high reputation'.[114] As to providing new work, Kropotkin was doubtful: 'I wish I could resume *Recent Science*. But it requires so much reading, thinking on thousands of details, and … under the present state of my health it is impossible. I have entirely given up the 12 or 14 hours' day work.'[115] While these lines confirmed the importance of *Recent Science* not only as paid employment but also as an intellectual endeavour, as discussed in recent studies,[116] the following lines also confirm that Kropotkin saw his ideas on mutual aid as intrinsically linked with the journal which had first published his views on this topic: 'In a month or two I shall take in hand *Ethics*, a continuation of the series in which Mr. Knowles was so interested, and which was begun in the Review, but interrupted by the first outbreak of revolution in Russia.'[117]

Although Kropotkin's work on *Ethics* would be finally resumed only during his later years,[118] he did collaborate with Skilbeck on a similar line of work, namely, the role of acquired characteristics in the theories of evolution and mutual aid, which formed the very foundations of ethics for Kropotkin. In a letter to his editor, the Russian anarchist wrote that, before going ahead with his project on ethics, he 'must discuss seriously the question of Darwinism, struggle for life and mutual aid. It is a big question as it requires a critical analysis of Natural Selection, but of the deepest interest just now, when Lamarckism is coming so prominently to the front. So, I have gone into it thoroughly, in the form of an analysis of the evolution of Darwin's ideas after the publication of the *Origin of Species,* as it appears from the 5 volumes of his *Letters'.*[119] Kropotkin's project involved a new series of consecutive articles similar to those he had published with Knowles, which was finally realised, including seven papers on mutual aid, the action of the environment on plants and animals, and the respective roles of inherited characteristics, published from 1910 to 1919 for *The Nineteenth Century.* [120] Kropotkin discussed with Skilbeck the growing extent of his work and the need for more articles, claiming that the field of investigation was ever increasing before his eyes. 'I never imagined that there might be such a mass of evidence in favour of the direct action of environment, never mentioned in several excellent recent books on Darwinism, and such a consensus of opinion in favour of the action of environment among the botanists.'[121] These comments reveal the substantial investment of the late Kropotkin in this field of investigation, which involved reconsidering works by the French early evolutionary theorist Jean-Baptiste Lamarck (1744–1829).

According to Alvaro Girón, this development arose due to an attempt to merge the ideas of Darwin and Lamarck in order to refute Malthusian ideas and to foster the scientific bases of cooperation. For Lamarck, the environment played a role in shaping inheritable characteristics relevant to evolution, and this viewpoint clearly reinforced the standard arguments of anarchist geographers that the benefits arising from mutual aid inform human and non-human beings in developing feelings of solidarity (including ethical feelings) in facing respective environmental challenges. According to Girón, these articles did not have the same scholarly success as the previous series on mutual aid, because such debates were already considered outdated within mainstream academia.[122] Nevertheless, Kropotkin's correspondence with Skilbeck shows that they remained a success in editorial terms. After receiving the second manuscript of the series, 'The direct action of environment on plants', the editor wrote to Kropotkin that: 'Your articles always add distinction to the Review and the sooner you can follow up this one with the next on the animals' response to their environment, the better I shall be pleased.'[123] These articles earned Kropotkin £40 each, slightly more than articles for the previous series.

While the tone of the exchanges between Kropotkin and his new editor always remained friendly, an incident occurred in 1910, on the occasion of a paper published by Ray Lankester (1847–1929) in *The Nineteenth Century,* on inherited characteristics and the environment.[124] A Darwinian biologist, a student of Huxley and former Oxford lecturer, Lankester raised serious doubts in respect of Kropotkin's competence in this field, which very much upset the Russian prince, who inundated Skilbeck with letters of complaint on this matter. Beyond a letter of 6 September 1910 requesting a meeting on these issues, the only surviving material from Kropotkin is a substantial number of minutes he kept in his archive, often erased and rewritten, which shows the intense concern Kropotkin had for this topic. First, he wrote to Skilbeck that he was 'painfully impressed by seeing that you had permitted Sir Ray Lankester to abuse me in this way on the pages of *The Nineteenth Century*. For more than twenty years I have been a scientific contributor to the Review. For ten years I had conducted – after Huxley had placed it in so high a [level] ... and, as Mr. Knowles wrote to me in 1897, none of my articles had ever required a single rectification'.[125] Rather than being personally angry, Kropotkin seemed to be more deeply saddened because he felt he had been somehow betrayed by a journal and an editor he considered as historical supporters of his ideas. Kropotkin insisted that he was not outraged because his ideas had been challenged, but rather because he considered some of Lankester's statements to be personally offensive, which accordingly aimed at undermining his scholarly and political positions: 'I have worded my letter (to Lankester) with great moderation. But I absolutely refuse to enter into any discussion, of any sort [with him], so long as he has not apologised.'[126] In the absence of a public apology, Kropotkin sought to win support from his editor, claiming that, if Lankester refused to apologise 'before your readers, before yourself and before me ... he is disqualified for being treated as an honest man'.[127]

Skilbeck found himself in a difficult position. First, he apologized to Kropotkin: '[I am] more than sorry that I should have published anything that should have given you a painful impression. I certainly did not imagine that it would do so, or it would not have been published.'[128] The editor promised to talk with Lankester: 'If you wish me to approach Sir Ray Lankester with regard to his withdrawing that statement ... I ought to do it soon.'[129] Beyond differences between their respective scientific views, Lankester's article contained statements overtly offensive to Kropotkin as a scholar. For example, the article began by declaring Kropotkin's 'complete misapprehension and consequent misrepresentation of the actual state of scientific knowledge',[130] due to his 'extraordinarily erroneous and misleading appreciation of the situation'.[131] Therefore, assuming Skilbeck's good faith in this affair, it seems clear that Knowles' successor was decidedly clumsier and less skilled in editorial diplomacy, or perhaps more careless, than his father-in-law, in allowing one of his contributors to publicly offend another. Skilbeck tried to reassure Kropotkin, arguing that: 'The expression of one man's opinion adverse to your work [cannot] have the effect which you suggest; your reputation is far too well established for that and, as you know, I was looking forward to your demolishing Sir Ray Lankester's arguments in a reply.'[132] After several exchanges, the correspondents agreed to deal with this problem in a footnote to Kropotkin's next article, which seemed to have closed this incident.

Whatever assessment one might give in relation to Kropotkin's attempts to reconcile Darwin and Lamarck, it is worth noting that recent biological scholarship is reconsidering Lamarckism and the possibility of inheritance of acquired characteristics in some organisms under certain conditions.[133]

A few surviving letters in the Moscow archives show that, with his publications in *The Nineteenth Century*, Kropotkin gained a very prestigious fan, namely, Alfred Russell Wallace, who had been one of the authors Kropotkin had read as part of his previous studies of evolution.[134] In 1889, Wallace invited Kropotkin to see him in Bournemouth, writing that he had 'read some of [his] articles in *The Nineteenth Century* and sympathise[d] greatly with [his] views'.[135] In that letter, Wallace expressed his political views, declaring his deep sympathies for socialism, although in its statist and not in its anarchist version:

> In social science, I have hitherto confined myself to the advocacy of the Land Nationalisation as the great reform needed in this country, and have believed that any system of complete socialism was impracticable in the present phase of human development. But that delightful and supremely important little work *Looking Backward* by Edward Bellamy of Boston-USA has convinced me that under a proper system of education and organisation there would be no insuperable difficulty in the working of a Socialistic State even now. In America something of the kind may come about soon. But here we must, I believe, pave the way for it by Land Nationalisation and other minor reforms.[136]

This letter confirms the important connections between science and socialism which existed in Victorian Britain. From the following letter, written by Wallace 13 years later, it is clear that, unlike in Reclus's case, Kropotkin and Wallace did not meet personally. Nevertheless, Wallace declared his appreciation for Kropotkin's 'very valuable work on *Mutual Help* [sic] of which I had the opportunity of reading in *The Nineteenth Century*'.[137] Wallace regretted that he had never met Kropotkin, 'as I am sure we have many opinions and sympathies in common ... With my best wishes for the success of your efforts in the cause of oppressed humanity'.[138] These declarations of sympathy confirm that Wallace was a dedicated reader of Kropotkin's work, as was already suggested in his published letters with other correspondents,[139] but these comments show in particular how *The Nineteenth Century* was important for Kropotkin's scholarly reputation and how evolutionary theory was a key factor in furthering relations between the sciences and various types of socialism in late Victorian and Edwardian Britain.

On the one hand, Kropotkin's exchanges with Knowles and Skilbeck show Kropotkin's determination in disseminating his work to the wider English-speaking public, while securing at the same time sufficient financial resources to sustain his life and work. On the other hand, they demonstrate the importance of sociability networks and of the personal relations formed within them. Kropotkin is revealed on several occasions to be in close personal involvement with his editors, who were simultaneously his friends, confidents and scientific advisors, as well as being involved in sometimes strongly-worded and deeply-felt disagreements on matters dear to Kropotkin's heart.

Kropotkin: an anarchist in the editorial business

Most of the articles Kropotkin wrote for the British periodical press, especially *The Nineteenth Century* and *Freedom*, constituted the bases for his most famous books such as *Mutual Aid, The Great French Revolution, The Conquest of Bread, Modern Science and Anarchism, Fields, Factories and Workshops*, and *Ethics*. The circulation of texts and content across different editorial support structures and languages, which included reworking, translations and re-translations from and into French, was complicated. As a general principle, already apparent in correspondence with Knowles, Kropotkin favoured maximizing the distribution of his texts for propaganda reasons through all possible outlets, while his publishers often tried to impose their conditions in terms of sales and rights of reproduction. The constant negotiation and attempts to balance these contrasting interests characterises the editorial history of Kropotkin's books.

These editorial discussions paralleled the negotiations carried on by Reclus in France with the big publisher Hachette, which were likewise marked by the author's willingness to circulate content to a wider public rather than solely to activists, and by the publisher's opening to unorthodox authors provided that their works were a commercial success. Kropotkin worked for Reclus and was paid directly by Hachette during his imprisonment in France,[140] corresponding directly with Reclus's cousin and Hachette cartographer Franz Schrader (1844–1924), who was a friend of all the anarchist geographers at that time, even though he did not hold radical political views. Kropotkin was also directly in touch with the London branch of Hachette in Charing Cross, which operated not only as a means to forward payments for Kropotkin's work on Reclus's *NGU* after he left prison,[141] but also to provide books and work-related material for the exiled scholar.[142] In the Hachette folder at GARF, two surviving documents show some ironical aspects of the anarchists' editorial collaborations. The first is a letter to Kropotkin from Schrader. Schrader, who was acquainted with Kropotkin through his cousins, complained about some of Kropotkin's articles on Siberia published in the *Geographical Journal* [143] because there, the Anarchist Prince had declared that maps completed by German cartographers from Gotha were the best on the region. Schrader protested that he had been one of the first to adopt Kropotkin's views on Siberian orography, in collaboration with David Aïtoff, who had been recommended to him by Kropotkin himself.[144] The national rivalry between the French geographical agency Hachette and the German publisher Perthes could not spare the anarchists! The second is a letter from Guillaume Breton, Hachette's associate and general director when the French publisher produced a translation of Kropotkin's *Mutual Aid*, asking for information about a request for copies from Paul Delesalle.[145] Delesalle (1870–1948) was an anarchist activist, a collaborator of Jean Grave for the journal *Temps Nouveaux*, and the fact that a prominent businessman served as an intermediary between two anarchists to facilitate their distribution of anarchist propaganda material is another ironic outcome of their editorial strategy involving the penetration of all possible milieus through scholarly work.

Aïtoff's name emerges again in editorial correspondence between Keltie and publisher McMillan concerning the *Statesman's Year Book*, a publication which Keltie edited from 1884 and for which he also requested contributions from Kropotkin. It is possible that Aïtoff, who was based in Paris and collaborated with publishers such as Hachette and Colin, had also been recommended to Keltie by Kropotkin. Giving work to friends and political bedfellows (Aïtoff was also a political refugee) was a defining characteristic of Reclus's and Kropotkin's editorial negotiations, when they could not take on work themselves. A letter from Keltie to Maurice McMillan provides an example of issues that sometimes subsequently arose in relation to the circulation of material, complicated by international negotiation and intellectual property issues:

The map showing the distribution of the races of Russia was sent to me by Mr. D. Aïtoff, who looks after the Russian section of the *Year Book*, for the revision of which he is paid each year. I told him I wanted a map showing the races of Russia and asked him to send me the materials. He sent me a copy of his own map to be used for the purpose, but never made any reference at all to Armand Colin. I took it for granted that Aïtoff had an absolute right to do what he liked with his own map. He had asked me to do so, I should certainly have written to Colin and added a note to the Map stating that it was published with his permission. I am writing to Aïtoff informing him about what Colin says and asking him to settle the matter with his publishers, as it appears to me that he is responsible for what has been done. At the same time, if it turns out that the Map is the absolute property of the Publisher, we cannot do anything in this year's issue, but he might promise that in the Preface or elsewhere in next year's issue he will make all due acknowledgement and I hope that Aïtoff will be able to resolve the matter amicably with Colin.[146]

As the *Year Book* was compiled through a collection of information coming from different countries, Keltie organised it following criteria similar to those that Reclus had adopted for the *NGU*, that is, through creating a centre of calculation[147] to collect information gathered as far as possible by correspondents situated in each country most likely to have first-hand information. Keltie claimed credit for this innovation during his conversations with McMillan, after more than 20 years of having edited this annual series: 'It took me some years when I first became Editor of the *Year-Book* to organize a system of getting information direct from the various states of the world. Our system is now completely organized and works efficiently.'[148]

As indicated in letters sent by Kropotkin, held at the British Library and in several GARF folders, Kropotkin started, in the 1890s, to employ the services of an international literary agency, run by George Herbert Perris (1866–1920) and Charles Francis Cazenove (1870–1915), who both engaged in extensive correspondence with Kropotkin, to manage the commercial and legal aspects of book publications. Running the 'Literary Agency of London' from 1899 to 1916, Perris was not only an agent and editor, but also a journalist, author of several books on the international political situation, especially interested in the Russian political refugees and an admirer of Count Tolstoy.[149] The business interests in common of Kropotkin and this agency need to be understood in the context of an editorial and writing history which saw 'a six-fold increase of those engaged in writing as their principal occupation'[150] in Britain between 1871 and 1911. Perris's agency was a successful commercial enterprise, and counted among its clients Patrick Geddes and a number of continental and Russian authors, including Tolstoy.[151]

Perris met Kropotkin sometimes at the Liberal Club in London, and their discussions concerned not only business matters, but also Russian politics.[152] His early correspondence with Kropotkin reveals that one of the key tasks of the agency was managing the transformation of Kropotkin's scholarly output into books palatable for the English-speaking public, which could include collections of articles already published elsewhere.[153] These letters indicate the outstanding success of Kropotkin's books in terms of sales and earnings. In 1902, Perris announced that 5,000 copies of *Fields, Factories and Workshops* (first published in 1899) had already been sold,[154] and a few weeks later he requested the author provide an updated preface for a new edition which he wished to print urgently.[155] Discussions on copyright were frequent, due to Kropotkin's wish for his work to be distributed widely among a varied reading public, beyond the considerations just of commercial return.[156] It is worth noting that Kropotkin's royalties were relatively rewarding; for example, he was offered by Chapman and Hall, for *The Conquest of Bread*, 'a royalty on 20% with £50 in advance'.[157] Despite this, the author was prepared to 'abandon entirely'[158] his rights of copyright whenever possible to allow a freer dissemination of his ideas.

In 1911, Kropotkin discussed with Cazenove an opportunity for selling the exclusive rights of *Fields, Factories and Workshops* to publisher Thomas Nelson, and expressed some concerns:

> The same uncertainty about the conditions offered by Nelson! The 100,000 copies to be sold means, plainly speaking, to sell the copyright of the book. But I discover in your letter of August 11, 1911 that you say: 'giving him the exclusive publishing rights for five years, OR until 100,000 copies have been sold'. Exclusive publishing rights for five years is quite right; it is a very fair clause. But does it mean that the exclusive rights run

beyond 5 years, if the 100,000 have been sold? At any rate, we must, I think, accept this proposal. As to the alternative Nelson suggested, in case of the book being published in a collected edition of my works, in which case the published price of such edition to be not less than 2£ per volume – The alternative, I think, so better drop altogether. Where it comes to such an edition, there will be time to settle the conditions. So, I think, we want simply to accept the proposal, and I have to be ready with the reviewed edition and the new seized?[159]

The book was finally published by Nelson in 1912, and the letter cited above highlights a number of important considerations in terms of Kropotkin's relations with the editorial business world. First, he was concerned to maintain a certain degree of freedom in deciding on how his texts were to be distributed, and his wariness therefore in being too strictly bound to one publisher, as happened with Reclus. Second, the public success of these writings allowed for various kinds of 'bargains'; in 1912, projecting 100,000 possible sales for an engaged book which had already been published in English a few years before, seems remarkable.

The same concerns were expressed by Kropotkin in relation to the publication of *The Conquest of Bread* by Chapman and Hall. This book, first published in French (1892) was partially anticipated in *Freedom* in the 1890s and then published in English as a volume in 1906, with a series of reprints over the following years. The 1913 edition provided an occasion to discuss the issue of copyright: 'Once Chapman and Hall intend to print 10,000 copies of the edition, I argue with you that this is already a certain guarantee. Will there be a clause about the [literary property] of the work to five [years] or after a time, as you have for *Fields Factories*?'.[160] A projected 10,000 copies for one paperback appears again as a remarkable situation. Astonishingly, Kropotkin did not consider Chapman as the best publisher for this type of edition, that is, the large-scale selling of cheap editions of his books, as shown by another letter discussing an offer Kropotkin received for his *Memoirs of a Revolutionist*:

I just receive a letter from Mr. Dent, the publisher. A friend of ours has spoken to him, saying that it would be well if he took my *Memoirs* into 'Everyman's Library', and Mr. Dent asks me to send him a copy of the book. I should not want to do that without first referring to you, as I understand that you are already in negotiations with Chapman and Hall on the same subject. To tell the truth, I should not much like to have the book in Chapman and Hall's hands, because they have not the necessary organisation for a large sale of cheap edition. With them I understand, it is a new venture, and nobody knows whether they will succeed. While the 'Everyman's Library' goes very well. The size of their books will also answer splendidly. Of course, Dent is not yet making a definite offer ... At any rate, I will not reply to him before I hear from you. Shall I not say that the matter is in your hands, and refer him to you?[161]

The Dent edition was apparently not realised, but the fact that, in 1913, Kropotkin was willing to leave everything in the hands of his agent is very telling. The editorial business activities he had started almost 40 years before when he arrived in Britain were finally successful, to the point that he was no longer able to manage their practical and commercial side, as he had done formerly with the help of his wife Sofia, who had also looked after the family finances at the time of Kropotkin's collaboration with Reclus.

Another issue arising with the publishers concerned the size of the books, involving the need to sometimes propose cuts, as well as how best to undertake image reproductions, as can be seen in respect of the projected Dent edition of the *Memoirs*, with Kropotkin writing again to Cazenove:

> Thank you very much for the trouble you take with the book. It is quite true that it is rather too big. But it would be impossible to take out so much as one third of it. I have not yet carefully considered the matter but I am afraid, it would be difficult to take out more than fifty pages without destroying the general character of the book. I am just looking over the American edition of the book and see that it has, all taken, with index, 520 pages, which may be reduced to, say, 480 pages of 350 words each. Of course, for a copyright book it may be too much; but otherwise the Everyman Series contains books from 350 to 500 pages, not to speak of the classics. As to give to Dent the plates, or moulds, we have none, I suppose: I do not believe that the Aberdeen press who have priced for us our last edition should have made moulds of it; and unless we buy the mould from the American publisher I do not see our way to supplying them. I am sending you the American edition that you might see how it looks. Only kindly return me this copy, because it is the only one of this edition which I possess. I am also sending you the dummy of a half-crown edition which the Aberdeen press had made for us when we were talking about our own edition. The dummy does not look too bulky.[162]

Beyond commercial considerations such as costs of production and author remuneration, Kropotkin was committed to keeping the price of his books low to facilitate public access to them. This caused Kropotkin to complain about Putnam, the American publisher of the *Conquest of Bread*:

> *Fields, Factories and Workshops*: in reply to your letter of November 9th, I quite agree with your proposal concerning the American edition. But what price will they put on it? Putnams, like all other publishers, find it hard to understand that my books find their readers in great numbers only when they are at a low price. Perhaps you will be more successful than I am in such cases, and will make them understand it. *Memoirs*: I have just compared the size of our Aberdeen page with Everyman's. Ours is larger than the usual Everyman page; but taking Shakespeare in Everyman's Library, I see that the height of our page is exactly equal to

theirs. ... So that if the publishers intended to make an edition of the *Memoirs* exactly like their edition of Shakespeare, they might do it very easily. ... It being so, we might make our own shilling edition. Swan Sonnenschein might do very well. The only difficulty is that it ought ... to be printed at the Aberdeen University Press, who employs non-University labour. Otherwise, I am sure they would make a good estimate for a 5,000 copies edition.[163]

Links of friendships likewise contributed to long-term collaboration between Kropotkin and his agents, especially Perris, who, interested in radical politics, once wrote to Kropotkin that he was joining a support campaign for Italian anarchist Amilcare Cipriani, another of Kropotkin's acquaintances in London for a time.[164] When Cazenove died prematurely in 1915, Perris sent Kropotkin acknowledgements from Cazenove's family for the anarchist prince's messages of condolence to them, and announced the *de facto* dissolution of his editing business, which accordingly put an end to their professional collaboration.[165]

Political agendas were still central to Kropotkin's publishing strategy even after having developed a niche in the mainstream book market. A fundamental aspect of these agendas involved the translation and multilingual circulation of ideas, an obvious concern for a transnational anarchist like Kropotkin. In a 1907 letter to publisher Heinemann, who had printed the first 1902 book edition of *Mutual Aid*, Kropotkin discussed terms for the translations of *The Great French Revolution*, which were proceeding alongside work on the English edition:

> The German, Dutch and Russian publishers took it without seeing the end, because they know my work, and the translations are going to be made ... to leave the English edition appearing at the same time as the others. ... Then, there is the main question; at what a price would you think it is possible to issue [the translation]?[166]

Kropotkin's discussions on price differences and differing uses of his works between countries account for the difficulty of these negotiations. In 1909, *The Great French Revolution* appeared concurrently in French, published by Stock in Paris, in English by Heinemann in London, in German by Thomas in Leipzig, in Spanish by the *Escuela Moderna* in Barcelona, in Dutch by the *Bibliotheek voor ontspanning en ontwikkeling* in Zandvoort, and in several other languages over the following years.[167] This highlights the links between anarchism and languages since, although Kropotkin had a certain success, according to Matthew Adams, in selling anarchism to the British,[168] most anarchist literature and material history until the last quarter of the twentieth century was first produced in languages other than English, especially (but not exclusively) in French, Spanish, Italian, German, Russian and Yiddish, as shown for instance by the complicated story of the multilingual translations of Reclus and Kropotkin.[169]

This situation needs to be taken into account for research, especially as studying anarchism has become more in vogue. A monolingual English-speaking scholar could not readily claim to be working effectively on anarchist histories, and scholarship should take as examples transnational scholars such as Reclus and Kropotkin who challenged linguistic barriers to conduct their internationalist programmes.

The business world of publishing again: Halford Mackinder and the anarchists

Among the acquaintances Reclus and Kropotkin shared within the milieus of the RGS, Gerry Kearns insists on the key role played by Halford Mackinder (1861–1947), which on the face of it is unusual, given Mackinder's substantially differing political views to those of the anarchist geographers.[170] According to Kearns, this phenomenon accounts for the political relevance of geography as a discipline, which facilitated discussion among scholars on fundamental political questions, for example, the use of state budgets for warfare or for education, in places which were formally considered neutral such as the RGS.[171] Kearns highlights the importance of 'traditions of academic and political tolerance'[172] which allowed the anarchist geographers to have a say in British learned societies and public arenas. While this is remarkable, material from the Moscow archives shows that the interests of the publishing business world also played a major role in these scholarly exchanges, and that Mackinder was mainly involved with Reclus and Kropotkin in relation to editorial commercial possibilities.[173]

When he met the anarchist geographers, Mackinder was completing a series of world regional geography books for the publisher Heinemann, in which Reclus was then subsequently engaged to take responsibility for Western Europe and Kropotkin for the Russian Empire. In their correspondence, the two anarchist geographers discussed how to align their respective contributions to the series. In January 1898, Reclus wrote to Kropotkin: 'You have sent me, with request to send it back, some quick notes about the plan of your book for the Mackinder collection: I used them to make sure our two books fit together. However, I hope we have the chance to meet and talk about that before work is finished'.[174]

The first documented invitation for publishing in this series was a letter Mackinder sent to Kropotkin in 1897, requesting a meeting in London to discuss the matter. Significantly, the proposed meeting place was the 'National Liberal Club'.[175] The choice of this institution, founded by Lord Gladstone to provide facilities for Liberal Party activists, suggests that Mackinder's sympathies for the Liberal viewpoint might have been considered by Kropotkin as an indicator that Mackinder's political ideas were not incompatible with his viewpoint, at least until the 1899 South-African war. Regardless, a subsequent letter reveals that this meeting could not take place due to Kropotkin's commitments and that the object of Mackinder's proposal was to make the Russian geographer an offer for an 'international geographical undertaking' in which 'Reclus, Markham, Keltie and others of authority are already cooperating'.[176] Notably, only one of the mentioned names would be included in the list of authors who actually published in that series.[177] It is also clear that Mackinder, still young and not yet well known, sought the authority and reputation of Kropotkin for business promotion. This need is also revealed by the humble tone of the letter, which ended with apologies for 'my lack of introduction. Had I remembered it this afternoon, I would have asked our mutual friend, Mr. Keltie, to give me a letter to you'.[178] In any case, the offer was accepted and Mackinder established a deadline for Kropotkin's manuscript, and even specified its size at 320 pages, for July 1899.[179]

In subsequent letters, other commitments for both men are shown to have become, at various times, reasons (or pretexts) for progressive delays, although Mackinder continued to remind Kropotkin of the money Heinemann was prepared to pay for the work, as well as mentioning new contributors such as Scottish geographer and map-maker John George Bartholomew (1860–1920), who was Geddes's and Reclus's collaborator and Kropotkin's acquaintance, as I explain below. However, the only concrete achievement was the publication of Mackinder's *Britain and British Seas* in 1902, which was sent to both Reclus and Kropotkin. The fact that, five years after the first contact, Mackinder was still acknowledging Kropotkin for his 'interest in the "Regions of the World" series' indicates some sort of problem with Kropotkin's engagement.[180]

In 1902, Reclus wrote again to Kropotkin, revealing some disappointment:

> What is the exact meaning of Mackinder's list of geographical books? The man announces my text as the first of the series, but I never received from him a precise and detailed letter ... I only receive, every year, one of these laconic postcards asking 'are you ready?'. I am not ready at all, albeit the work would advance if I had data, numbers, definitive instructions.[181]

In 1903, the three men had occasion to meet during Reclus's presentation of his spherical atlases at the RGS. Later, Mackinder wrote his last surviving letter to Kropotkin, lamenting that:

> When we talked together at the Reclus meeting of the RGS, I gathered that your volume on Russia for the Regions of the World series had not progressed. Mr. Heinemann urges that if that is so, we should do well to consider the matter closed. If I judge your feeling currently you should not be sorry, but I should like to hear from you on the subject.[182]

In the end, Reclus's and Kropotkin's volumes were never completed. According to Kearns, 'there clearly was a problem with this series and with Mackinder's management of it',[183] also related to possible editorial issues, as the last titles were not published by Heinemann, but by Clarendon Press. However, the newly available sources quoted above allow for several further observations. First, Mackinder's interest in Reclus and Kropotkin was not only due to intellectual curiosity and liberal intellectual tolerance, but also for reasons linked to editorial business promotion. Mackinder had been a young geographer when Reclus and Kropotkin had already become well known international authorities, and it was beneficial to secure and publicly display their involvement as part of his series publications. Second, despite Mackinder warmly recognising the anarchist geographers, welcoming Reclus at the RGS as 'our brother geographer,'[184] the anarchist geographers seemed to have maintained a relatively cool attitude towards him. Their collaboration declined, and apparently ended in the early years of the twentieth century, exactly at the time when the South-African war had exacerbated nationalism, jingoism and opposition from Liberal quarters. Therefore, the political distance between them noticeably increased. Kropotkin's correspondence with Keltie continued until 1917, but direct contact with Mackinder seems to have stopped in 1903. This suggests that tolerance and open-mindedness from RGS members were important to the anarchist geographers, but also that the anarchists were not mere passive recipients of solidarity, and pursued a conscious political and scholarly strategy which included the selection of which partners to best work with.

The intermingling of politics, scholarship and editorial business, helps explain much of the British and international networks of Reclus and Kropotkin.

Notes

1   Ferretti, 'Publishing anarchism'.
2   Adams, *Kropotkin*, 27.
3   Ferretti, *Élisée Reclus*.
4   Fyfe, 'Journals, learned societies and money'; Ogborn and Withers, *Geographies of the Book*.
5   Mayhew, 'Materialistic hermeneutics'.
6   Ferretti, 'Publishing anarchism'; Kearns, *Geopolitics*; Kropotkin, *Memoirs*.
7   Chabard, 'Un elephant blanc'; Chabard, 'Towers and globes'.
8   National Library of Scotland (hereafter NLS), Ms 10522, Kropotkin to Geddes, 13 September 1882.
9   Reclus, *Correspondance, vol. II*, 267.
10 Apparently, the other exile was the former nihilist Nikolai Vasilyevich Tchaikovsky (1850–1926).
11 NLS, Manuscript 10522, Keltie to Geddes, 27 December 1882.
12 NLS, Manuscript 10522, Keltie to Geddes, 8 February 1883.
13 NLS, Manuscript 10522, Keltie to Geddes, 1 March 1883.
14 Hyndman, *The Record*, 261–262.
15 Ferretti, 'The correspondence'.
16 GARF, 129, op 2, khr 2653, Hutt to Kropotkin, 30 March 1883; 18 June 1883.
17 Reclus, *HT, vol. VI*, 3–5.
18 Allen, *Joseph Cowen*.

19 Miller, *Kropotkin*, 155.
20 GARF, 1129, 2, 1428, Cowen to Kropotkin, 25 July 1881.
21 GARF, 1129, 2, 1428, Cowen to Kropotkin, 23 September 1881.
22 GARF, 1129, 2, 1428, Cowen to Kropotkin, 10 August 1881.
23 Hug, *Peter Kropotkin*, 103–105.
24 GARF, 1129, 2, 1428, Cowen to Kropotkin, 22 November 1881.
25 Miller, *Kropotkin*, 156.
26 GARF, 1129, 2, 1428, Cowen to Kropotkin, 12 December 1881.
27 GARF, 1129, 2, 1428, Cowen to Kropotkin, 7 and 27 March 1882.
28 GARF, 1129, 2, 1428, Cowen to Kropotkin, 22 June 1882.
29 GARF, 1129, 2, 1428, Cowen to Kropotkin, 20 July 1882.
30 GARF, 1129, 2, 1428, Cowen to Sofia Kropotkin, 26 December 1882.
31 GARF, 1129, 2, 1428, Cowen to Sofia Kropotkin, 29 December 1882.
32 Avakumović and Woodcock, *The Anarchist Prince*, 189.
33 GARF, 1129, 2, 1428, Cowen to Westall, 23 February 1883.
34 GARF, 1129, 2, 1428, Cowen to Westall, 23 February 1883.
35 Ferretti, *Élisée Reclus*.
36 Cahm, *Kropotkin*; Miller, *Kropotkin*.
37 GARF, 1129, 2, 1428, Cowen to Kropotkin, 25 June 1883.
38 GARF, 1129, 2, 1428, Cowen to Sofia Kropotkin, 7 February 1884.
39 GARF, 1129, 2, 1428, Cowen to Sofia Kropotkin, 1 June 1886.
40 GARF, 1129, 2, 1428, Cowen to Sofia Kropotkin, 26 March 1889.
41 Also in touch with Reclus, Keane was the translator of part of the English edition of the *New Universal Geography*.
42 GARF, 1129, 2, 1308, Keltie to Kropotkin, 15 September and 31 October 1881.
43 Livingstone, 'Finding revelation'.
44 Hug, *Peter Kropotkin*, 105–115.
45 GARF, 1129, 2, 2148, Robertson Smith to Kropotkin, January 29 [1889].
46 Mavor, *My Windows*, 79.
47 GARF, 1129, 2, 2148, Robertson Smith to Kropotkin, 16 January [1889].
48 GARF, 1129, 2, 1308, Keltie to Kropotkin, 12 January 1894.
49 Ferretti, *Élisée Reclus*.
50 GARF, 1129, 2, 613, Black to Kropotkin, 29 August 1882.
51 GARF, 1129, 2, 613, Black to Kropotkin, 19 June 1884.
52 GARF, 1129, 2, 613, Black to Kropotkin, no date.
53 GARF, 1129, 2, 1309, Scott Keltie to Kropotkin, 9 August 1899.
54 GARF, 1129, 2, 1309, Scott Keltie to Kropotkin, 5 September 1899.
55 GARF, 1129, 2, 1309, Scott Keltie to Kropotkin, 23 July 1900; GARF, 1129, 2, 1309, Scott Keltie to Kropotkin, 31 October 1900.
56 GARF 1129, 2, 2785, Chisholm to Kropotkin, 11 July 1905.
57 GARF 1129, 2, 183, Kropotkin to Chisholm, 22 November 1907.
58 GARF 1129, 2, 183, Chisholm to Kropotkin, 25 November 1907.
59 Brun, 'Introduction'.
60 Kinna, *Kropotkin*; Varengo, *Pagine anarchiche*.
61 Ferretti, 'Organisation'.
62 GARF 1129, 2, 183, Kropotkin to Chisholm, no date, draft letter.
63 GARF 1129, 2, 183, Kropotkin to Chisholm, no date, draft letter.
64 Dickenson, 'The naturalist'.
65 Kropotkin, *Memoirs, vol. II*, 318.
66 Kearns, *Geopolitics*, 86.
67 Dickenson, 'The naturalist', 210.
68 GARF, 1229, 2, khr 546, Bates to Kropotkin, 20 September 1882.
69 GARF, 1229, 2, khr 546, Bates to Kropotkin, 6 October 1882.
70 Driver, *Geography Militant*; Kearns, *Geopolitics*.

71   GARF, 1229, 2, 546, Bates to Kropotkin, 9 December 1882.
72   Avakumović and Woodcock, *The Anarchist Prince*, 228.
73   Pedgley, 'Mill'.
74   Kearns, *Geopolitics*, 86.
75   Mill and Freshfield, 'Obituary'.
76   Mill, *An Autobiography.*
77   Royal Geographical Society with the Institute of British Geographers [hereafter RGS], HRM, 3, Kropotkin to Mill, 12 May 1892.
78   RGS, HRM 3, Kropotkin to Mill, 13 November 1893.
79   RGS, HRM 3, Kropotkin to Mill, 19 October 1893.
80   RGS, HRM 3, Kropotkin to Mill, 3 March 1893.
81   RGS, HRM 3, Kropotkin to Mill, 6 March 1895.
82   RGS, HRM 3, Kropotkin to Mill, 14 November 1894.
83   Ferretti, *Élisée Reclus.*
84   Withers, Finnegan and Higgit, 'Geography's other histories?'
85   GARF, 1129, op. 2, khr 1754, Mill to Kropotkin, 1 November 1898.
86   GARF, 1129, op. 2, khr 1754, Mill to Kropotkin, 15 March 1900.
87   RGS, HRM 3, Kropotkin to Mill, 9 March 1900.
88   Ferretti, 'La première agence'.
89   RGS, HRM 3, Kropotkin to Mill, 28 January 1901.
90   Ferretti, 'The correspondence'.
91   GARF, op 2, khr 1754, Mill to Kropotkin, 17 July and 3 October 1895.
92   RGS, HRM 5, Reclus to Mill, 2 April 1895.
93   RGS, HRM 5, Reclus to Mill, 29 November 1895.
94   RGS, HRM 5, Reclus to Mill, 29 November 1895.
95   The collaboration between Kropotkin and Knowles is further discussed in Ferretti, 'Publishing anarchism'.
96   Ferretti, 'Publishing anarchism'
97   Metcalf, *James Knowles*, 324, 327.
98   GARF, 1129, 2, 1895, Knowles to Kropotkin, 14 April 1885 and 11 February 1888.
99   Kropotkin, *Fields, Factories and Workshops.*
100   GARF, 1129, 2, 1895, Knowles to Kropotkin, 20 March and 9 July 1896.
101   Metcalf, *James Knowles*, 285.
102   GARF, 1129, 2, 1308, Keltie to Kropotkin, 11 March 1881.
103   GARF, 1129, 2, 1895, Knowles to Kropotkin, 19 January 1889.
104   Secord, *Victorian Sensation*, 261–296.
105   GARF, 1129, 2, 1895, Knowles to Kropotkin, 13 November 1905.
106   GARF, 1129, 2, 1895, Knowles to Kropotkin, 25 March 1906.
107   City of Westminster Archives Centre [hereafter CWAC], 716/84, Kropotkin to Knowles, 27 March 1906.
108   CWAC, 716/84, Kropotkin to Knowles, 30 March 1906.
109   GARF, 1129, 2, 1895, Knowles to Kropotkin, 2 April 1906.
110   CWAC, 716/84, Kropotkin to Lady Knowles, 18 February 1908 and Kropotkin to Skilbeck, 10 July 1908.
111   GARF, 1129, 2, 2276, Skilbeck to Kropotkin, 13 July 1908.
112   Purchase, *Peter Kropotkin.*
113   GARF, 1129, 2, 2276, Skilbeck to Kropotkin, 17 July 1908.
114   CWAC, 716/84, Kropotkin to Skilbeck, 20 July 1908.
115   CWAC, 716/84, Kropotkin to Skilbeck, 20 July 1908.
116   Ferretti, 'Publishing anarchism'.
117   CWAC, 716/84, Kropotkin to Skilbeck, 20 July 1908.
118   Kropotkin, *Ethics.*
119   CWAC, 716/84, Kropotkin to Skilbeck, 16 November 1909.

120 Hug, *Peter Kropotkin*.
121 CWAC, 716/84, Kropotkin to Skilbeck, 14 April 1910.
122 Girón, 'Kropotkin Between Lamarck and Darwin'.
123 GARF, 1129, 2, 2276, Skilbeck to Kropotkin, 18 July 1910.
124 Lankester, 'Heredity'.
125 GARF 1129, 2, 148, Kropotkin to Skilbeck, no date.
126 GARF 1129, 2, 148, Kropotkin to Skilbeck, no date.
127 GARF 1129, 2, 148, Kropotkin to Skilbeck, no date.
128 GARF, 1129, 2, 2276, Skilbeck to Kropotkin, 15 September 1910.
129 GARF, 1129, 2, 2276, Skilbeck to Kropotkin, 14 September 1910.
130 Lankester, 'Heredity', 423.
131 Lankester, 'Heredity', 424.
132 GARF, 1129, 2, 2276, Skilbeck to Kropotkin, 15 September 1910.
133 Wang, Liu and Sun, 'Lamarck rises'.
134 Hale, *Political Descent*, 227.
135 GARF, 1129, 2, 2525, Wallace to Kropotkin, 2 September 1889.
136 GARF, 1129, 2, 2525, Wallace to Kropotkin, 2 September 1889.
137 GARF, 1129, 2, 2525, Wallace to Kropotkin, 19 October 1902.
138 GARF, 1129, 2, 2525, Wallace to Kropotkin, 19 October 1902.
139 Fichman, *An Elusive Victorian*; Marchant, *Alfred Russel Wallace*.
140 Ferretti, 'The correspondence'.
141 GARF 1129 2, 405, [illegible signature] to Kropotkin, 9 November 1887.
142 GARF 1129 2, 405, Hachette & C to Kropotkin, 1 February 1898. GARF 1129
      2, 405, 25 September 1901; GARF 1129 2, 405, 15 June 1906.
143 Kropotkin, 'Orography of Asia'.
144 GARF 1129, 2, 405; Schrader to Kropotkin, 31 August 1904.
145 GARF 1129, 2, 405; Breton to Kropotkin, 28 June 1906.
146 British Library, Ms Add 5542, Keltie to McMillan, 7 June 1906.
147 Latour, *Science in Action*.
148 British Library, Ms Add 5542, Keltie to McMillan, 27 May, 1907
149 Gomme, *George Herbert Perris*.
150 Gomme, *George Herbert Perris*, 150.
151 Gomme, *George Herbert Perris*, 153.
152 GARF, 1129, 2, 1980, Perris to Kropotkin, 8 January 1896.
153 GARF, 1129, 2, 1980, Perris to Kropotkin, 28 November 1892.
154 GARF, 1129, 2, 1980, Perris to Kropotkin, 19 February 1902.
155 GARF, 1129, 2, 1980, Perris to Kropotkin, 17 March 1902.
156 GARF, 1129, 2, 1980, Perris to Kropotkin, 15 July 1905.
157 GARF, 1129, 2, 1980, Perris to Kropotkin, 23 August 1906.
158 GARF, 1129, 2, 1257, Cazenove to Kropotkin, 29 October 1906.
159 British Library, RP 7435, Kropotkin to Cazenove, 6 December 1911.
160 British Library, RP 7435, Kropotkin to Cazenove, 10 July 1912.
161 British Library, RP 7435, Kropotkin to Cazenove, 26 October 1912.
162 British Library, RP 7435, Kropotkin to Cazenove, 2 November 1912.
163 British Library, RP 7435, Kropotkin to Cazenove, 19 November 1912.
164 GARF, 1129, 2, 1980, Perris to Kropotkin, July 21 [no year].
165 GARF, 1129, 2, 1980, Perris to Kropotkin, 20 June 1915.
166 GARF, 1129, 2, 32, Kropotkin to Heinemann, 26 March 1907.
167 Hug, *Peter Kropotkin*, 86–89.
168 Adams, *Kropotkin*, 27.
169 Errani, Élisée Reclus, geografia sociale; Hug, *Peter Kropotkin*.
170 Kearns, *Geopolitics*.
171 Kearns, 'The political pivot'.
172 Kearns, 'The political pivot', 345.

173 I especially thank Pascale Siegrist, who first shared with me copies of the Mackinder folder surviving at GARF.
174 GARF, 1129, 2, 2103, Reclus to Kropotkin, 13 January 1898.
175 GARF, 1129, 2, 1646, Mackinder to Kropotkin, 26 July 1897.
176 GARF, 1129, 2, 1646, Mackinder to Kropotkin, 7 December 1897.
177 Kearns, *Geopolitics*, 184.
178 GARF, 1129, 2, 1646, Mackinder to Kropotkin, 7 December 1897.
179 GARF, 1129, 2, 1646, Mackinder to Kropotkin, 26 August 1898.
180 GARF, 1129, 2, 1646, Mackinder to Kropotkin, 27 February 1902.
181 Reclus, *Correspondance, vol. III*, 284.
182 GARF, 1129, 2, 1646, Mackinder to Kropotkin, July 1903.
183 Kearns, *Geopolitics*, 184.
184 Reclus *et al.*, 'On spherical maps and reliefs. Discussion', 294.

3   Establishing a geographical tradition in the 'British Isles'

Emergent social and political geographies

This chapter addresses some of the scholarly outputs of these editorial networks, especially Reclus' *Guide* for the (French) traveller in London, the chapter of Reclus's *NGU* on the British Isles, first published in French and then translated into English under the supervision of Ernst Ravenstein,[1] and Kropotkin's landmark work *Fields, Factories and Workshops*, an account of the British economy, particularly the English economy. This latter work had been developed after Kropotkin's observations of the spatial patterns of capital and industries in Britain, which greatly influenced the emergence of anarchist theory concerning economic and productive decentralisation, which inspired later English-speaking anarchist thinkers such as Murray Bookchin and Colin Ward. I also consider that Reclus's and Kropotkin's biographical and intellectual contacts within the British Isles were crucial for developing some of their main contributions to both geography and anarchy, providing therefore a further demonstration of the effects that places can exert on the production of knowledge. This precise effect of a specific place on the development of ideas is evident with an early elaboration of the concept of social geography by Reclus, and with what can be described as Kropotkin's economic geography that served simultaneously as an alternative to Marxist economy-centrism and was instrumental in introducing economic problematics into the anarchist debates. As is shown below, geography and anarchism were closely interconnected in the works, thoughts and places of these early scholars and activists.

The analyses undertaken by the anarchist geographers provided an original viewpoint on British economic histories and their spatialities. With his descriptions of pauperism in the industrial cities, Reclus questioned what has been considered the 'Victorian Boom' from the 1850s and 1870s, showing how capitalist growth tended to marginalise certain people, especially in otherwise well-developed urban and rural regions, anticipating contemporary understandings of 'uneven development'.[2] The relevance of Reclus's remarks on issues related to sanitation and urban mortality has been confirmed by contemporary studies in British historical demography, which have shown that only in the 1870s did mortality rates start to fall, initiating a process of demographic transition.[3] While Reclus's early writings on England had involved witnessing outcomes of the first industrial revolution, later writings by the anarchist geographers, especially Kropotkin's *Fields, Factories and Workshops*, took into account the second industrial revolution to argue for the possible social uses of technology to foster social progress. Ideas resulting from these works would inspire in turn Patrick Geddes's concepts of the paleotechnic and the neotechnic as technologies associated with new possibilities of social progress.[4]

However, a significant characteristic of the second industrial revolution, about which Kropotkin provides early analysis, was the progressive loss of the British monopoly on manufacturing, confirmed by economic historians who have observed that, 'in the fifty years before the First World War, Britain lost the unchallenged position which it had gained as the first industrial nation'.[5] Kropotkin addressed this phenomenon to question Smithian and Ricardian commonly accepted views on the needs for specialisation and on developing regionally specific productive capacities that would lead to the creation of a 'core' and a 'periphery' in terms of world production during the twentieth century, later criticised by the dependency theorists. The effectiveness of Kropotkin's analysis lay in denaturalising the relations of production beyond historical and geographical determinism, highlighting their character as involving political and cultural choices. Establishing foundations for emerging social and economic geographies, these early writing of Reclus and Kropotkin contributed to generating an anarchist and critical geographical tradition through analyses based on conclusions drawn from the British Isles.

A universal geography, and an amazing traveller's guide

According to Gary Dunbar, the earliest occurrences of the term 'social geography' can be traced to French literature in relation to works such as those of Reclus in the *NGU* from 1884, and in *L'Homme et la Terre* later.[6] Nonetheless, in studying the milieus of the anarchist geographers, it is possible to detect some earlier occurrences of the term in anarchist debates which took place in Switzerland, within the *Fédération jurassienne*, in the 1870s. In 1877, according to Max Nettlau, Kropotkin proposed *Social Geography* as the title for the second part of a projected collective book *Geographical Sketches* targeting popular education, in a debate which involved Élisée Reclus and James Guillaume. While the description 'social geography' as proposed by Kropotkin was very comprehensive, it was, however, very telling. According to Kropotkin, this work, 'which can only be done and published by socialists, would include: an introduction, a summary on the formation of societies, the making of property, the free communes, the states and the present-day monopolising and exploiting state. Then, a geography describing countries, their resources and natural treasures, what happens now with them and what might happen tomorrow'.[7] In a nutshell, 'social geography' was understood as a geography explicitly claiming social and political relevance.

Therefore, it is unsurprising that Reclus's *NGU* was described as 'Studies of Social Geography' by French sociologist Paul De Rousiers (1957–1934), a follower of Frédéric Le Play (1806–1882), whose school likewise exerted a strong influence on young Patrick Geddes.[8] As with many of his fellow sociologists, De Rousiers was very interested in the relations between geographical factors and human societies, and considered Reclus's work as an important resource:

> Noticing that Switzerland is covered by mountains and that most of Hannover is a wide plain might be of course useful information for the travellers who are willing to visit these countries, but it is not sufficient. On the contrary [Reclus] analysed features of work, family, property, salary, ruling classes and their auxiliary institutions on these different soils, so that his work assumes a scientific character and belongs to the range of documents to consult to study human societies.[9]

De Rousiers articulated a type of geographical determinism where 'the place determines the work, and all social phenomena of property, family and patronage join these two primary elements as corollaries'.[10] Although Reclus's views on this point were much more complex and nuanced,[11] the definition of social geography was nevertheless adopted explicitly by the anarchist geographer to introduce *L'Homme et la Terre*. According to Reclus: 'Class struggle, quest for balance and sovereign decision of the individual are the three main kinds of facts which are revealed to us by the study of social geography and, in the general chaos of things, they prove to occur in sufficiently regular ways to call them "laws"'.[12]

Showing London to the French

Some of these ideas on social geography can already be found in one of Reclus's earliest work, namely his 1860 *Guide to London*, which was very different from similar works published at that time. Indeed, it not only contained insights on how to visit the main monuments of the city, but also included chapters on poverty, unhealthy living conditions and social problems. After richly illustrated pages describing the London world fair and a first chapter of general information, the second chapter introduced issues such as pauperism, criminality, hygiene and education to which several dozen pages were dedicated. After describing the magnificence of the British capital for his French readers, Reclus claimed that all his readers needed to be informed of the darker side of this 'world city', that is, all that 'this society rejected'.[13] In very explicit terms, Reclus stated that: 'One cannot deny that the spectacle offered by the slums of the English society are more repugnant than everything else exists of lamentable and painful on Earth. The more English society improves its glory and power, the deeper its roots sink into the muddy soil.'[14] This biological metaphor clearly recalled related socialist arguments that great wealth was not the solution to great poverty but rather its cause. The primary cause of this type of development was identified as liberal economics: 'Economic laws, which increasingly give to England an incontestable political primacy on the world, concurrently sank and relegated in the mud a part of its population.'[15]

As with many who focussed on issues of hygiene at that time, Reclus attributed a lack of light and open air as responsible for a great part of the diseases which affected poor urban neighbourhoods. With an impressive rhetoric style, sometimes difficult to translate from French, the anarchist geographer compared the destiny of London's slum dwellers to those working in 'Welsh and Scottish mines, who come into the world, grow and die in a thick atmosphere, charged of mortal gaz. Their only horizon are the columns of smoke which appear and disappear at the lamps' light, and their only sky is an awful ceiling under which an eternal despair flows.'[16] After his experience in Ireland, Reclus had considered the Irish peasant as the stereotype of the very poor person, but, after having become familiar with English slums and industrial cities, he argued that the rural poor at least enjoyed a better environment, while the urban poor starved in the most polluted and dirtiest possible places. For Reclus, this showed again 'how much horror, suffering and bitterness in the big city's slum is needed for allowing grand lords and grand ladies reaching such level of sumptuousness and comfort'.[17]

A chapter of his guide was devoted to issues of urban health and sanitation, and he discussed statistics indicating 10,000 deaths each year linked to diseases associated with poor housing sanitation in London, as well as almost 100,000 unemployed in the capital city of the most industrialised nation in the world at the time. The chilling accounts which Reclus offered to his 'traveller' of the spectacle of everyday misery in London's streets recalled novels of British authors like Charles Dickens, who was also explicitly quoted by the French anarchist geographer. Precarious sanity conditions were likewise associated with a reduced life expectancy in neighbourhoods such as Whitechapel, and in the case of prostitutes, 'whose average lifetime hardly exceeds 26 years'[18] and whose number was estimated to range from 8,600 to 80,000 throughout London. In any case, Reclus highlighted the enormous extent of that particular social problem, as well as issues to do with crime and of adulterated food, as shown in official statistics. The French geographer also condemned the practice of locking poor people up in hospices and workhouses, anticipating his later critiques of prisons and of what is called today 'total institutions'.[19]

The solutions Reclus proposed were essentially social and socialist: 'To resolve these problems and give to all the city the healthiness which is now enjoyed only by the privileged neighbourhoods of the West End, there would be only one efficient solution: the abolition of pauperism. However, waiting for the realisation of such ideal, London's municipality and the Parliament could make plenty of things to sanitize the city.'[20] First, this should involve replacing existing hovels with 'shelters worthy of the munificence of the richest community in the world'.[21] The details of his descriptions of hovels included matters related to fire risk and of sewer management. Reclus's depressing descriptions of the lives of sewer-scavengers[22] recalled contemporary French realistic literary descriptions of Paris's poor, written in the same years by Victor Hugo, or in the 'Paris mysteries' by Eugène Sue.[23] For Reclus, the worst conditions existed in neighbourhoods still devoid of sewers, with health statistics used to focus on this social problem: 'According to Mr. Southwood Smith, if we take a map of London on which we indicate the districts affected by fevers, and we compare them with a map of the city's sewers, we can be sure that diseases always find their cradle in the districts which are not yet sanitized with sewers.'[24] A map of London's sewers would be reproduced in the *NGU* to support the same conclusion (Figure 3.1).

As discussed in the *NGU*, London and the British Isles were among the first laboratories for Reclus's application of his concept of social geography. The type of travel guides he produced was doubtlessly unusual not only for that time but also for more recent periods. It is noteworthy, for example, that, two years later, Hachette published an abridged and 'illustrated' edition of

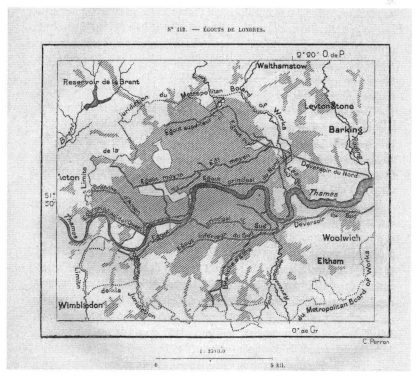

Figure 3.1 Égouts de Londres [London's sewers]
NGU, vol. IV, 505

Reclus's *Guide to London*, at half the size of the first edition, and with the chapters devoted to social problems excluded.[25]

The *NGU* and the Mediterranean metaphor

Reclus's *NGU* was considered a masterpiece and a classic of 'modern' geography. This work was translated into several languages. Concerning the English language edition, recent scholarship has revealed a degree of 'infidelity' in the translated version supervised by Ravenstein, especially concerning Ireland, where Reclus's sentences blaming British colonialism were simply deleted.[26] Nevertheless, Ravenstein had been one of Kropotkin and Reclus's British acquaintances, and had been acknowledged among other contributors who had furnished insights for Reclus's work, together with Henri Woodward and others.[27] I quote here excerpts from the chapters on Britain taken from the original French due to the numerous examples of political 'unfaithfulness' in translations of Reclus's books in several languages,[28] as well to allow for a consideration of this work as an example of what Le Play's followers defined as social and political geography which reflected the anarchist ideas of its author. In the *NGU*, Reclus's geographical analyses of the British Isles highlighted the importance of maritime communications, including British rivers and estuaries, which were shorter than on the continent but constituted some of the most important communication networks on earth: 'The Thames is not the largest of British rivers, but the historical importance of its basin is comparable to few others in the world.'[29] Similar descriptions were extended to the Severn estuary, and to the Forth-Clyde system, which rendered Glasgow 'the first European city to be regularly reached by steamboats'.[30]

The city of London was compared to Paris in relation to its favourable position alongside an advantageous waterway, which secured at the same time rapid access to the sea and ready defence from sea-borne attacks. Reclus in his discussion also nuanced and relativized geomorphological determinism, using history and arguing that the locational advantages accruing to settlements depended considerably on the technology characterising each epoch. According to Reclus, London began to exploit its locational advantages internationally only after the English state had attained political unity and stability. On a global scale, London was described as lying 'at the centre of the continental hemisphere'.[31] The description 'continental hemisphere' recalls Carl Ritter's classes attended and summarized by Reclus,[32] where general principles concerning possible correlations between human history and the shaping of the landscape were introduced. Two main concepts developed by Ritter included the Mediterranean metaphor and the principle of coastal indentation. The first concept consisted in a progressive understanding of the oceans as 'internal seas', given that the concept of an internal sea as a means for historical communications and exchanges between opposite shores had been extended by technological innovations such as steamboats and telegraphs.[33] For Ritter in particular, the Atlantic Ocean was at the centre of what he called the 'continental hemisphere', which corresponded to approximately half of the globe occupied by the continental masses of Asia, Africa, Europe and the Americas, opposed to the 'oceanic hemisphere' mainly occupied by the Pacific and Indian Oceans. Therefore, the British Isles enjoyed an especially favourable position in the Atlantic for developing commercial routes, an idea of Ritter's that was extended by Reclus's analyses of the maritime world activity of the city of London: 'London is not far from the geometrical centre of all the continental masses. Therefore, no other city in the world is better placed to become the convergence port of all the navigation lines.'[34] Reclus also expressed a strong interest in London's cosmopolitism, which he understood as the presence of a multi-ethnic, not exclusively white population, matching his ideas of universal brotherhood and generalised miscegenation as an antidote against racism.[35] With some humour, Reclus claimed that, in London's docks, 'one meets hundreds of sailors of all the world's races, Blacks, Hindus, Malayans, Chinese, Polynesians: nowhere else on earth could one undertake so complete ethnologic studies'.[36]

London also provided one of the first opportunities for Reclus to present his ideas on modern urbanisation and urban growth. The directions of urban growth were described through organic and naturalistic metaphors, for example, being compared to an octopus (later repeated by Patrick Geddes),[37] or the roots of a tree, or a lake breaching its banks. These ideas were expressed in graphic form by Reclus's cartographer, Charles Perron, who drew maps of London's historical development (Figure 3.2) and of its surface area compared to that of Paris, to depict the extraordinary demographic importance of the British capital at that time (Figure 3.3).

London's environs were also the subject of one of the coloured maps included in the *NGU* as separate sheets (Figure 3.4). Reclus's letters to Hachette's editorial director Charles Schiffer reveal the 'dynamic' of these documents. Without going into the technical details, it is worth noting that this document, drawn by French map-maker Alexandre Vuillemin, caused Reclus some headaches, because the growth of the city was apparently happening faster than the updating of the maps used as sources by Vuillemin, yet it was precisely the growing dimensions of London and its urban areas that Reclus wanted to convey accurately in this map:

> London is far bigger, and the interest of this map is exactly in showing its enormous dimensions. The map shows that Hampstead, Highgate and all the villages [that I listed] can be comprised under the same red colour. I am sending you a map which is two years old, but far more updated than Vuillemin's first attempt. I think he used an old version of the Ordnance Survey map.[38]

As Reclus's archetype of what we might call today a global or world-city,[39] London was also presented as providing a frightening series of examples on the social and environmental evils afflicting modern industrial cities. As with

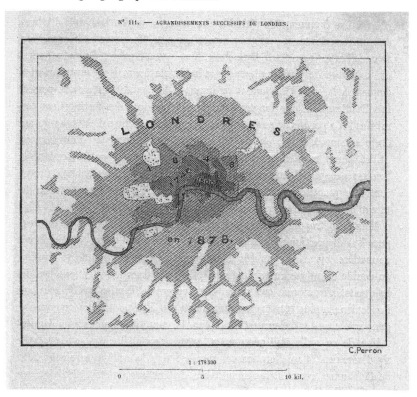

Figure 3.2 Agrandissements successifs de Londres [London's historical development] *NGU*, vol. IV, 503

his *Guide*, Reclus discussed serious issues of water management and health in regard to heavily populated areas, which had apparently improved little over the previous 15 years: 'Unluckily, the English capital does not provide sufficient clean water: many neighbourhoods only receive a liquid replete with organic materials in decomposition. By way of consequence, mortality rates rise at double or triple the rate of those in the neighbourhoods where inhabitants drink clean water.'[40] London served as a key case study in trying to understand the complexities of urbanization which characterised Reclus's thinking (and equally that of Kropotkin and Geddes). On the one hand, it was apparent that cities kill people. According to Reclus, the proof of this observation was that, in most industrial cities, demographic growth was only ensured by migration from the countryside, as deaths exceeded births, although in this respect London was a fortunate exception because there 'births exceed deaths'.[41] On the other hand, cities were places of culture, sociability and social movements. Therefore, according to the anarchist geographers, the advantages of the city and those of the countryside should be integrated, as I explain below.

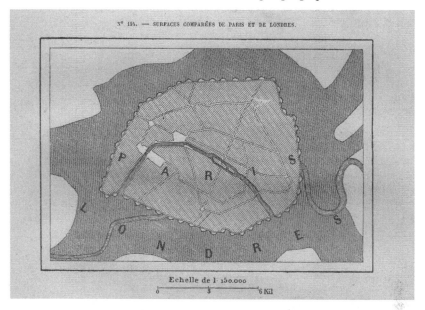

Figure 3.3 Surfaces comparées de Paris et de Londres [Surface area comparisons of
        Paris and London]
NGU, vol. II, 718

Cities were additionally places where class contradictions were most on
display and spatialised. Reclus's analyses anticipated some features of the
contemporary concept of gentrification: 'In the last 40 years, something
like 50,000 workers have been driven out from the City and have accumu-
lated in the periphery. There are poor people no longer in the City, but
they are more and more numerous in the suburbs ... What a contrast
between the look of poor neighbourhoods and that of the sumptuous
ones.'[42] In Reclus's description of London, there can also be found a
statement of the importance of social investigation into underprivileged
areas, as opposed to the banality of a simple description of the most
famous monuments, which Reclus considered as symbols of evil, arguing
that: 'The Tower of London's history is that of royal crimes'.[43] Class contradictions also affected cultural institutions that Reclus greatly admired,
such as the art museum in South Kensington: 'It is sad that such an
institution, which is a true school of art and science, is at the extremity of
one of the aristocratic neighbourhoods of London, far from the city centre
and from the population it should benefit.'[44] Another influential principle
that the anarchist geographers developed for studying urban developments
was the necessity for understanding cities in terms of the effects of their
surrounding territories acting at different scales; for London, the proper
contexts were both England and the British empire.

Figure 3.4 Londres et l'estuaire de la Tamise [London, and the estuary of the Thames]
NGU, vol. IV, 500

The *NGU* and the principle of coastal indentation

The second principle established by Ritter and adopted by Reclus concerned the role of coastal indentations as a possible driver for historical developments. This principle has been recently rediscovered by authors widely criticised in geographers' milieus, and linked especially with environmental determinism, eventually Jared Diamond and Robert D. Kaplan.[45] However, it is worth noting that forms of environmental determinism are returning in a number of scholarly works,[46] which suggests it may be important to avoid simplistic readings of categories such as determinism. Authors such as Franco Farinelli have already demonstrated that the alleged opposition between 'possibilism' and 'determinism' in this field was an invention of the French historical *Annales* school, and any Manichean distinction between the two categories does not allow for their complexity, especially when one is determined as 'good' and the other as 'evil'.[47]

The first application of the principle of coastal indentation occurred in relation to early navigation in the Mediterranean, especially in the ancient Greek and Phoenician worlds, as an example of civilisations connected by early travel and maritime communications. While drawing parallels between European colonial powers and the 'splendour' of Greek civilisation was a commonplace of imperial ages in restating European centrality,[48] Reclus used the comparison between ancient Greece and the modern British Isles as a device for his dynamic idea of geo-history.[49] In Reclus's view, environmental 'influences' were relativized in relation to different historical situations and the availability of different technologies. According to Reclus: 'The remarkable analogy that the small Greek peninsula and the close islands present with the archipelago of Great-Britain, lying at the other side of the continent, is repeated by their respective commercial roles. Albeit in a different environment and in a different historical cycle, the same natural advantages brought similar results.'[50] These advantages for politics and trade lay not only in the presence of estuarine and coastal specificities, but in their relative position. For the islands of Britain, being located relatively close to the continent was undoubtedly advantageous in terms of being within a reasonable distance to trade with other European countries and being relatively well insulated against continental armies. Furthermore, in the period of early globalisation, the British Isles were relatively well-placed in relation to the main Atlantic trade routes. According to Reclus, while the British had not been the first to circumnavigate the globe or the first to open maritime routes between Europe and Eastern or Western 'Indies', the British Isles were nevertheless able to extend political power or cultural influence over a seventh of the world's surface area, despite having dimensions that comprised only one six-hundredth part of that area.

Although Reclus's critiques of colonialism and colonial crimes have been highlighted in recent scholarship,[51] the anarchist geographer considered the British Isles as the cradle of another aspect that came to characterise the

British Empire, namely, the capitalist system resulting from the first industrial revolution. Reclus understood the two phenomena as interlinked. As British and Irish sailors 'contributed to discover and appropriate unknown lands, miners also took possession of another world, that of subterraneous wealth'.[52] Therefore, the link between oceanic connections and the availability of raw material such as coal fostered the spread of industry from cities such as Manchester, Sheffield, Birmingham and Leeds. As with other social critics of that time such as Morris or Geddes, Reclus was clearly aware of the negative consequences arising from this production of wealth. First, he stressed the environmental evils occurring in relation to the transformation of agricultural villages and ancient walled towns into industrial cities 'where the sky is always blackened by coal; also, the most sumptuous houses are stained with soot; black snow falls on trees and meadows'.[53] Second, the social question had emerged as an issue on which Britain pioneered international developments, specifically, the workers' movement. As I explain below, criticisms of British 'civilisation' were even sharper in *L'Homme et la Terre*, although the British Isles were also praised by Reclus for facilitating the workers' movement of his day through the welcoming of refugees (albeit not disinterested), and the birth of the First International in 1864: 'It was on British soil, and with the participation of English brothers, that occurred the founding of different groups of social and political solidarity, from which the decisive movement of the International sprang.'[54]

For Reclus, industrialisation 'transformed the social state, created a new lifestyle, preparing great revolutions for humankind. In England there occurred first the terrible problems of the modern proletariat. It is there that the most numerous human masses depend on stock market movements, where strikes are the biggest and where workers associations have most force'.[55] Reclus's critique of British society focused on its deep social inequalities. 'A hideous misery weighs upon millions of English people; social inequality, harsher in England than in all Latin countries, has dug an abyss between the rich and the poor, between the landlord and the rural worker, between the master and the servant.'[56] Given this situation and the lack of social mobility, it was not surprising that hard-fought social struggles took place in the coal sector, where the likelihood of strikes was exacerbated by 'the diminution of wages … and the dreadful methane explosions which can cause the death of hundreds of miners'.[57] In the textile sector, Reclus noted with admiration 'the most numerous workers' strike in modern history',[58] which took place in Preston from 1853 to 1854. High worker mortality rates were condemned as resulting from excessive periods of work and the overexploitation of women and children, and he noted that, despite formal legal protection, 'the rate of children who are already disabled at their thirteenth birthday is still increasing'.[59] Therefore, a conclusion that the *NGU* was a work pioneering the definition and use of 'social geography' is more than justified, despite some French interpretations considering the *NGU* as a more conventional work in comparison with Reclus's other writings.[60]

A critical point for this new approach of social geography was the way in which Reclus addressed the problem of languages. In the *NGU* volume on the British Isles, written at the end of the 1870s, Reclus considered English as a possible universal language, due mainly to its demographic strength in terms of speakers and the geographical spread of those speaking it, which Spanish and French could not match. In 1872, he wrote to Richard Heath apologising: 'Sorry for my errors in English: we are far from reaching the universal language'.[61] In the *NGU*, he argued that English was also suitable for becoming a universal language because it belonged to two linguistic families: 'The German group for its origin, spirit and phrasing; the Latin one for its lexical variety'.[62] He also thought that the relative ease of its syntax and grammar would also help in its general adoption. It might seem surprising to see a famous anarchist from continental Europe praising English, which is considered today as the imperial language par excellence. Reclus's arguments, and the further developments of his thinking, help clarify his position. Reclus posed the following question: 'Might it not be possible to extend instruction to all the world's countries, and that the study of a language serving to foster relations between peoples be added to the study of the maternal language that represents a people's own genius and their treasure of national thinking and aspirations?'.[63] Although the idea that an extension of primary public education could also involve an extension of bilingualism may appear advanced for the nineteenth century (and beyond), in the last volume of *L'Homme et la Terre*, published in 1908 after Reclus's death, there is reference made to Esperanto, an international language only recently invented at that time. With the rise of Esperanto, Reclus considered its adoption as a universal language to be generally preferable to existing languages such as English or French, given their inability to pacify 'national rivalries',[64] as well as possibly fostering what he had always desired, namely, universal brotherhood, 'a feeling which now mainly characterises socialist workers'.[65] It is worth noting that, in Reclus's depiction of the world's 'commercial languages',[66] the two which occupied more than half the number of all speakers were indicated as Chinese and Hindi. Therefore, it seems that the idea of a common language was primarily a pragmatic matter for Reclus. Whether Esperanto, English or some other language, the urgent task was to put people into mutual communication despite political, linguistic and cultural barriers, to foster internationalism and worker solidarity, and also to promote intellectual cosmopolitanism against the narrowness of nationally based scholarship. In June 1908, *Freedom* published a report in favour of Esperanto, presented at the International Anarchist Congress which had taken place in Amsterdam the year before.[67]

However, a defence of languages threatened with extinction was also a topic of Reclus's geographies, including Scottish, Welsh and Irish Gaelic. In the *NGU*, the peninsulas of Cornwall and Wales were compared to France's Brittany in terms of cultural proximity and language, with the Welsh language defined as 'closer to Armorican and to ancient Cornish than to Scottish or Irish, [though] it has survived better than Breton and has an incomparably richer literature'.[68] We find here some anticipation of the idea of the 'Celtic Fringe'[69] and an appreciation of the survival capacity of linguistic and cultural minorities in mountainous areas, (Wales and Scotland in this case), later addressed by left-libertarian authors including James C. Scott.[70] In Scotland, Reclus identified two main regions which roughly corresponded to a plain/mountain dichotomy that also characterised much geographical literature during the Romantic period, namely, the urbanised and industrial area between the isthmus of Solway and Clyde-Forth, and the Highlands, regarded as socially and politically distinct since Roman times, and where 250,000 people still spoke Gaelic, despite invasions and mixing among 'Saxons, Angles, Bretons ... Scots, Frisians, Danish and Norwegian Scandinavians'.[71]

The Irish question in Reclus's work has been studied extensively in recent work.[72] In the following chapter, an account on how the Irish issue was dealt with in anti-colonial terms in the journal *Freedom* provides an important example of the type of ideas circulating in the activist milieus to which Reclus and Kropotkin directly contributed in the British Isles.

Fields, factories and workshops: an anarchist economic geography of England

Both Reclus and Kropotkin supported an integration of town and countryside, and this idea was also reflected in their residential choices, often oriented, where they had a choice, towards areas which one might consider today as 'peri-urban'.[73] Indeed, their relation to cities was ambivalent. On the one hand, they considered cities as pestiferous places filled with social and sanitary problems caused through industrial pauperism, which correspond to contemporary ideas in 'urbaphobics'.[74] On the other hand, they considered cities as necessary places for human sociability and intellectual activity, and saw modern transportation, especially trains and electric tramways, as a way for enjoying at the same time the rest of the countryside and the cultural stimulations of cities through what we might call today 'commuting'. In his residential choices in England, Kropotkin preferred to keep at a distance from London, settling first in Harrow on the Hill, relatively close to the city, from 1886 to 1891, then moving steadily further away, to Acton (1891–1894), then to Bromley, Kent (1894–1907), to Highgate (1907–1911) and finally to Brighton (also for health reasons), where he remained until his departure for Russia in 1917.[75]

These ideas of Reclus and Kropotkin on decentralisation exerted a significant influence on urban thinkers such as Patrick Geddes and Lewis Mumford, and those associated with the garden city movement, possibly Ebenezer Howard.[76] In this section, I claim that Kropotkin's ideas were also the result of observations carried out in the British Isles, especially in England, leading to his concept of productive decentralisation, which influenced thinking on urban matters throughout his networks. The more 'British' of

Kropotkin's works, *Fields Factories and Workshop*, which contained an original and spatial appreciation of the English economy in comparison with other countries, established key principles for anarchist thinking, such as decentralisation and labour integration. *Fields Factories and Workshops* can also be considered as an important book of economic geography, which challenged Marxist assumptions on the increasing concentration of capital (whose capacity for spatial extension was empirically examined by Kropotkin). This work can still inform debates within economic geography on globalisation as a phenomenon intrinsically linked with a plurality of localisations,[77] and on non-monetary economies, which contemporary literature has identified as demonstrating creative bases on which mechanisms of mutual aid are daily created within current society.[78]

This book, first published in 1898, was also the result of Kropotkin's collaboration with three of his most important British acquaintances: James Knowles (as discussed in Chapter 2), Edward Carpenter, and James Mavor (discussed in Chapter 5). Kropotkin's work started with a challenge to well-known laws, first articulated by Adam Smith, on the need for a division of labour to promote specialisation, both to increase trade between nations and to increase individual production. In these areas, the first consideration Kropotkin raised was not social class, but rather 'humanity', in whose name he argued that 'there is no advantage for the community in riveting a human being for all his life to a given spot [without] the highest enjoyments of science and art, of free work and creation'.[79] For Kropotkin, the division of labour produced increasing brutalisation among the working classes and was not the result of economic laws, but of a contingent historic situation. In fact, it was the result of the industrial revolution supported by initial British protectionism and naval power, which had in turn determined Britain's monopoly of advanced industrial activities. Nevertheless, for Kropotkin, 'the monopoly of industrial production could not remain with England for ever. Neither industrial knowledge nor enterprise could be kept for ever as a privilege of these islands. Necessarily, fatally, they began to cross the Channel and spread over the Continent.'[80] Kropotkin observed a newly emerging increase of international competition in the context of what was later called the second industrial revolution, and concluded that the protection of national industry was proving to be more and more ineffective within international economic scenarios, where 'the decentralisation of manufactures goes on with or without protective duties — I should even say, notwithstanding the protective duties... the monopoly of the first comers in the industrial field has ceased to exist'.[81]

Kropotkin analysed the moral aspects of this situation, noting critically how the dictates of the free market favoured one nation at a certain moment and its competitors in different historical periods. For Kropotkin, if 'capital knows no fatherland, [then] capital will migrate to India, as it has gone to Russia, although its migration may mean starvation for Lancashire and Dundee'.[82] This was the kind of chaos implied by liberal capitalistic markets against which Reclus famously considered anarchy as 'the highest expression of order'.[83] In the name of a new possible order, aimed at removing pauperism and capitalist distortions, and working towards an economy based on real needs, Kropotkin challenged the alleged necessity of ensuring British exports, considering that real progress involved 'producing for home use'.[84] Kropotkin's primary argument paid tribute to the capacity of peasants and workers from Britain and Ireland to be the principal suppliers for their own local demand: 'Why the British worker, whose industrial capacities are so highly praised in political speeches; why the Scotch crofter and the Irish peasant, whose obstinate labours in creating new productive soil out of peat bogs are occasionally so much spoken of, are no customers to the Lancashire weavers, the Sheffield cutlers and the Northumbrian and Welsh pitmen?'.[85] Production for local consumption was then presented as a key possibility for consideration.

Kropotkin first analysed the possibilities arising within agriculture, where the author targeted beliefs on the 'naturalness' of famines and of the alleged poverty of some lands in relation to richer land elsewhere that served as a foundation for believing that land use should 'naturally' involve specialised commodity production. For Kropotkin, the question was: 'Is it possible that the soil of Great Britain, which at present yields food for one-third only of its inhabitants, could provide all the necessary amount and variety of food for 41,000,000 human beings?'[86] While current opinion at the time stated the impossibility of such self-sufficiency in food, Kropotkin argued that empirical observations radically contradicted this commonly held belief. Kropotkin considered the reason for the relatively low productivity of English pasture land was not due to natural infertility, but the 'abandonment of the land'[87] caused by the enclosures and by historical migration towards the towns and cities.

During his excursions into rural areas around London, Kropotkin noticed the lack of human occupational activity on land that appeared almost like a desert in areas which could be used for supplying the city: 'within ten miles from Charing Cross, close to a city with 5,000,000 inhabitants' land use could ensure the city was 'supplied with Flemish and Jersey potatoes, French salads and Canadian apples'.[88] An alternative approach was evident in the Parisian *culture maraîchère*, a traditional practice of urban gardening which still worked in supplying local markets, and which showed increasing rates of productivity due to the use of more advanced technologies for manuring and heating the soil. Kropotkin discussed intensive agriculture as an example of land use, and of using greenhouses and techniques for improving the productivity of single plots of land through examples drawn from the United States. For Kropotkin, the powerful international competitiveness of North American agricultural markets was not only due to the fertility and vast extent of the land involved, but also to strong investments in technologies which allowed North Americans to take the lead in 'intensive, or forced, agriculture as well'.[89] Therefore, productivity was not a 'natural issue', but a matter involving technologies and choices.

Kropotkin's other forcefully articulated argument in relation to agriculture concerned an attack on the laws of Thomas Malthus (1766–1834), who had famously argued that there was an inevitable disproportion between population growth and resources, which meant poverty was an inescapable feature of the natural world. Kropotkin observed that Malthus's *Essay on the Principle of Population* (1798) was primarily a reaction against the ideal of equality raised by the French Revolution and especially by William Godwin's proto-anarchist ideas:

> It was precisely when the ideas of equality and liberty, awakened by the French and American revolutions, were still permeating the minds of the poor, while the richer classes had become tired of their amateur excursions into the same domains, that Malthus came to assert, in reply to Godwin, that no equality is possible; that the poverty of the many is not due to institutions, but is a natural *law*. [90]

To ridicule this alleged law, Kropotkin attacked two of its pillars. First, the idea of limited supply, or scarcity, and second, the idea that the amount of population was indirectly proportional to the possibility of wealth for everybody. Against this, Kropotkin argued that, despite economists' explanations, the real world of peasants showed that a dense population providing intensive labour can have a multiplier effect on field productivity. Empirical observations were increasingly demonstrating that no absolute limit could be put to the potential productive capacity of any single acre of land, since 'our means of obtaining from the soil whatever we want, under *any* climate and upon *any* soil, have lately been improved at such a rate that we cannot foresee yet what is the limit of productivity [which] vanishes in proportion to our better study of the subject'.[91] This anti-deterministic argument with its appeal to inventive possibilities for enhancing production allowed Kropotkin to make a remark appealing to British patriotism, that 'the greenhouse for commercial purposes is essentially of British, or perhaps Scottish, origin'.[92] Finally, Kropotkin's data on technological innovations in agriculture allowed him to argue empirically against Malthusian understandings of over-population, as 'it is precisely in the most densely populated parts of the world that agriculture has lately made such strides'.[93]

Kropotkin's book subsequently addressed the problems of manufacturing in Britain and other industrialised countries, contending that a commonly assumed opposition between 'the two sister arts' agriculture and industry was basically an artificial distinction, given that, in history, the integration of rural and manufacturing practices had not been unusual. Kropotkin aimed to discuss the possibilities for this integration to occur in Britain at that time, a study which necessarily started with an appreciation of 'rural industries, domestic trades and petty trades'.[94] As a result of his observations in Britain, Kropotkin found that, despite the increasing growth of big industries, 'the numbers of those who earn their livelihood in the petty trades most probably equal, if they do not surpass, the numbers of those employed in the factories.[95] In excursions he made with Edward Carpenter in Sheffield, and in other observations he conducted in Leeds, Birmingham and London, Kropotkin could appreciate how the old spirit of craftmanship was surviving in a myriad of small trades and activities being undertaken alongside the major industries, and how they were an important factor for British industrialisation. Concerning the spatiality involved for the exercise of these crafts, Kropotkin noticed their tendency to agglomerate in towns, which anticipated the contemporary concept of an industrial district. For Kropotkin, former activities dispersed throughout the land and initially destroyed by industrialisation tended to be spontaneously reconstructed in urban contexts, 'side by side with the great factories', which he considered not a peculiarity, but an 'economic necessity'.[96] In fact, for Kropotkin, new industries constantly created new needs, pushing many factories to do what today might be called an 'outsourcing' of some parts of their production. To do that, and to satisfy new needs which urban and industrial life entailed, new 'petty trades' surged into life. Finally, Kropotkin considered these innovations as an indicator of the 'inventive genius of a nation',[97] which anticipated many contemporary debates concerning the economy in relation to creativity.

Nevertheless, it would be wrong to conclude that Kropotkin was focusing on small-scale production while neglecting large-scale complex processes. His claim for a 'combination of agriculture with industry'[98] was intended as a concrete proposal for transforming the economy and society of what was then the most advanced imperial and economic power of the world. Kropotkin's programme was not therefore simply reformist, urging a return to the land, but a project involving an integral transformation of society and its spaces. This entailed challenging what the Russian geographer called 'leviathan factories',[99] with implicit and not coincidental reference to Thomas Hobbes, and directly confronting the economic organisation of capitalism together with its spatial reproduction, exemplified by the slums of British industrial cities. In contrast, producing at 'the starting point' meant producing not for profit but 'for the producers themselves',[100] and required an integral transformation of society, one which demanded not only the elimination of economic exploitation, but also 'the necessity for each healthy man and woman to spend a part of their lives in manual work in the free air ... Humanity as a whole, as well as each separate individual, will be gainers by the change, and the change will take place'.[101]

This process of social transformation would have as an outcome the possibility of realising an integration of manual and intellectual work for each individual, challenging social hierarchies in the context of a society where people, 'with the work of their own hands and intelligence, and by the aid of the machinery already invented and to be invented, should themselves create all imaginable riches'.[102] What Kropotkin proposed was an early model of a genuinely prosperous society, which had rejected the dogmas of the market and of scarcity, and which focussed on human needs and humane possibilities for everybody in relation to well-being, health and happiness beyond what is now known as econometric indicators.

A classic work of anarchist thinking, *Fields Factories and Workshops* did not fail to exert a degree of influence within British culture, and 20 years after the book's first publication, Bertrand Russell could claim that: 'Returning from [Malthus's] dim speculations to the facts set forth by Kropotkin, we find it proved in his writings that, by methods of intensive cultivation, which are already in actual operation, the amount of food produced on a given area can be increased far beyond anything that most uninformed persons suppose possible'.[103] In conclusion, the British Isles offered multiple sorts of inspirations for early anarchist geographers. British and Irish places, cultures, authors and traditions played roles to varying degrees in the development of some of Reclus's and Kropotkin's fundamental ideas in matters in relation to geographies of poverty, anti-colonialism, migrations, and industrial decentralisation. Conversely, the anarchist geographers contributed to greater understanding of British and Irish geographies through spreading information on these regions among continental readerships and within radical networks.

Notes

1 Ferretti, 'Political geographies'.
2 Smith, *Uneven Development*.
3 Baines and Woods, 'Population and regional development'.
4 Geddes, *Cities in Evolution*.
5 Floud, 'Britain', 1.
6 Dunbar, 'Some early occurrences'.
7 Nettlau, *Eliseo Reclus, vol. II*, 50.
8 Fowle and Thomson, *Patrick Geddes*.
9 De Rousiers, 'Etudes', 511.
10 De Rousiers, 'Etudes', 514.
11 Clark and Martin, *Anarchy, Geography, Modernity*; Ferretti, 'Evolution and revolution'.
12 Reclus, *HT, vol. I*, 4.
13 Reclus, *Guide du voyageur*, 90.
14 Reclus, *Guide du voyageur*, 90.
15 Reclus, *Guide du voyageur*, 90.
16 Reclus, *Guide du voyageur*, 91.
17 Reclus, *Guide du voyageur*, 91.
18 Reclus, *Guide du voyageur*, 98.
19 Reclus, *HT, vol. VI*.
20 Reclus, *Guide du voyageur*, 103.
21 Reclus, *Guide du voyageur*, 104.
22 Reclus, *Guide du voyageur*, 116–117.
23 Hugo, *Les misérables*; Sue, *Les mystères de Paris*.
24 Reclus, *Guide du voyageur*, 118.
25 Reclus, *Londres illustré*.
26 Ferretti, 'Political geography'.
27 Reclus, *NGU, vol. IV*, 961.
28 Ferretti, 'Traduire Reclus'.
29 Reclus, *NGU, vol. IV*, 481.

30 Reclus, *NGU, vol. IV*, 663
31 Reclus, *NGU, vol. IV*, 501.
32 Reclus, 'Introduction'
33 Arrault, 'A propos du concept de Méditerranée'.
34 Reclus, *NGU, vol. IV*, 356–357.
35 Ferretti. 'Revolutions, and their places'.
36 Reclus, *NGU, vol. IV*, 511.
37 Geddes, *Cities in Evolution*.
38 BPUN, MS 1991/10, Reclus to Schiffer, 31 May 1878.
39 Massey, *World City*.
40 Reclus, *NGU, vol. IV*, 504.
41 Reclus, *NGU, vol. IV*, 508.
42 Reclus, *NGU, vol. IV*, 516–517.
43 Reclus, *NGU, vol. IV*, 520.
44 Reclus, *NGU, vol. IV*, 531.
45 Diamond, *Guns, Germs and Steel*; Kaplan, *The Revenge of Geography*.
46 Livingstone, 'Changing climate'.
47 Farinelli, 'Come Lucien Febvre'.
48 Dussel, 'Europe'.
49 Ferretti, 'Anarchism, geo-history'.
50 Reclus, *NGU, vol. I*, 56.
51 Ferretti, 'They have the right'.
52 Reclus, *NGU, vol. IV*, 357.
53 Reclus, *NGU, vol. IV*, 358.
54 Reclus, *HT, vol. VI*, 22.
55 Reclus, *NGU, vol. IV*, 358.
56 Reclus, *NGU, vol. IV*, 369–370.
57 Reclus, *NGU, vol. IV*, 835.
58 Reclus, *NGU, vol. IV*, 615.
59 Reclus, *NGU, vol. IV*, 840.
60 Lacoste, 'Élisée Reclus'.
61 Reclus, *Correspondance, vol. II*, 88
62 Reclus, *NGU, vol. IV*, 361.
63 Reclus, *NGU, vol. IV*, 363.
64 Reclus, *HT, vol. VI*, 466.
65 Reclus, *HT, vol. VI*, 467.
66 Reclus, *HT, vol. VI*, 467.
67 E. Chapelier and G. Marin, 'Anarchists and the international language, Esperanto'. *Freedom*, June 1908, 28; July 1908, 56.
68 Reclus, *HT, vol. VI*, 389.
69 Hechter, *Internal Colonialism*.
70 Scott, *The Art of not Being Governed*.
71 Reclus, *HT, vol. VI*, 655.
72 Ferretti, 'Political geography'.
73 Homobono, 'Las ciudades'; Oyón and Serra, 'Las casas'
74 Baubérot and Bourillon, *Urbaphobie*.
75 Avakumović and Woodcock, *The Anarchist Prince*.
76 Boardman, *Patrick Geddes*; Ferretti, 'Situated knowledge'; Pesce, *Da ieri*.
77 Dicken, *Global Shift*.
78 White, 'Explaining'; White and Williams, 'The pervasive nature'.
79 Kropotkin, *Fields*, 3.
80 Kropotkin, *Fields*, 9.
81 Kropotkin, *Fields*, 11.
82 Kropotkin, *Fields*, 27.

83 IISH, Élisée Reclus Papers, ARCH01170, Développement de la liberté dans le monde, 1851.
84 Kropotkin, *Fields*, 38.
85 Kropotkin, *Fields*, 38–39.
86 Kropotkin, *Fields*, 43.
87 Kropotkin, *Fields*, 47.
88 Kropotkin, *Fields*, 48.
89 Kropotkin, *Fields*, 82.
90 Kropotkin, *Fields*, 83.
91 Kropotkin, *Fields*, 103.
92 Kropotkin, *Fields*, 112.
93 Kropotkin, *Fields*, 120.
94 Kropotkin, *Fields*, 127.
95 Kropotkin, *Fields*, 134.
96 Kropotkin, *Fields*, 139.
97 Kropotkin, *Fields*, 146.
98 Kropotkin, *Fields*, 177.
99 Kropotkin, *Fields*, 180.
100 Kropotkin, *Fields*, 183.
101 Kropotkin, *Fields*, 183.
102 Kropotkin, *Fields*, 219.
103 Russell, *Roads to Freedom*, 102.

4 Striving for *Freedom*

Reclus's and Kropotkin's politics in the UK

This chapter examines the journal *Freedom*, founded in 1886 by Kropotkin and Charlotte Wilson, as one of the centres of Reclus's and Kropotkin's sociability networks in the British and Irish world. Drawing on invaluable work on *Freedom* by Heiner Becker, and more recently by Selva Varengo,[1] and on previously mentioned literature discussing London in its role as the principal nucleus for transnational anarchist networks at that time,[2] I present here the results of a systematic survey of the first 32 years of *Freedom*, from 1886 to Kropotkin's departure in 1917 (although as noted his rupture with the editorial group had begun in 1914). In addition, an extensive examination of the surviving correspondence between Kropotkin and the key journal editors and contributors was undertaken. My main claim is that *Freedom*, in like manner to the corresponding French series *Le Révolté-La Révolte-Les Temps Nouveaux* edited by Jean Grave,[3] provided a model expression of the anarchist-communist line of thinking then represented internationally by Reclus and Kropotkin, and engaged with their respective scholarly work, especially Kropotkin's in the case of *Freedom*. In the expression of political viewpoints, geography was considered as an important source of inspiration and information, nourishing anarchist critiques of colonialism and overseas wars, and informing struggles for internationalism and cosmopolitanism.

Freedom also importantly exemplified two other widespread characteristics of anarchism as a political movement: its openness and pluralism. Anarchism needs to be understood in its networks and in its capacity to influence different actors and milieus rather than in the number of organised and self-declared anarchists, which despite some luminous exceptions remained generally limited. As a historian of anarchist women and Emma Goldman's biographer Kathy E. Ferguson commented: 'Anarchism spread on the surface of communities, moving along their capillaries, circulating within their discourses. Like feminism, anarchism enabled a resonance that exceeded its specific parts, an interactive energy that touched many relationships and shaped many events'.[4] The links between *Freedom* and the wider British socialist networks that I analyse below help to substantiate this claim, including Ferguson's comparison between anarchism and feminism. Anarchist networks can be situated in the complex and pluralist traditions of early British socialism, influenced not only by religious dissent and by the French Revolution, but also by locally specific phenomena such as the Levellers and Diggers, including figures such as Gerrard Winstanley (1609–1676), deemed a Kropotkin forerunner by Jim MacLaughlin and something of a libertarian by Marie-Louise Berneri.[5] This confirms the importance of tracing the (situated) social networks of scholars and activists. As James Hulse has observed, London was likewise the place where Karl Marx and Friedrich Engels settled for most of their lives in England, and Marx's residences in London remain objects of veneration for his admirers, but 'Marx was a recluse'[6] and, likewise according to Hulse, Engels also had little contact with the English working classes. Kropotkin's highly developed capacity for networking with activists and scholars of different tendencies was a function both of the possibilities opened up by the discipline of geography, and of the less dogmatic attitude of anarchists in dealing with British socialist traditions in comparison with most Marxists of that time.

Indeed, this chapter on *Freedom*, together with the following chapter on the 'Humane Sciences', shows that the milieus of the anarchists based in Britain and of their closer political allies in the socialist field, were particularly sensitive to topics such as anti-colonialism (including the Irish question), anti-racism, women's emancipation, same-sex love, civic rights and political liberties, vegetarianism and animal rights, which are topics often neglected in traditional anarchist historiography but well represented in the works and activities of early anarchists, and especially the anarchist geographers, anticipating contemporary directions in social change suggested by concepts such as intersectionality.

Charlotte and Pyotr: founding a journal

At the time that Kropotkin was released from jail and returned to London in January 1886, three main groups characterised the socialist scene in Britain: the Social-Democratic Federation led by Hyndman, akin to Marxism but, according to Kropotkin, taking 'a revolutionary attitude'[7] in the years 1887–1889; the Socialist League, which held anti-parliamentary positions inspired by Morris and others, and the Fabians, considered as more moderate and dedicated to cultural work, represented by figures such as George Bernard Shaw (1856–1950). Anarchist activity was increasing in those years, and Kropotkin started collaboration with a group editing *The Anarchist*, led by Henry Seymour (1861–1938), an anarchist of individualist tendencies who published 40 issues of this journal from March 1885 to April 1888. The force of Kropotkin's personality was the prime mover in causing a definitive split between communist and individualist tendencies. According to Bantman, Kropotkin 'joined the group of *The Anarchist*, which in April was renamed *The Anarchist, Communist and Revolutionary*. Within a few months, however, communists and individualists were clashing, and Wilson and Kropotkin left to launch their own paper',[8] *Freedom, a Journal of Anarchist Communism* [hereafter *Freedom*]. In a tract written by Charlotte Wilson, 'A brief history of *Freedom*', the foundation act is presented in very general terms: 'The first number of *Freedom* was started in October 1886, by Peter Kropotkin and C.M. Wilson, the latter acting as Editor'.[9] The founding of the journal is further clarified in Kropotkin's recollections in an unpublished letter to Alfred Marsh, stating: 'In March 1886, Seymour came to ask me to work with him. I replied with cold words, as his *Anarchist* was individualist … We agreed all to work on his *Anarchist*, but it did not work and we founded *Freedom*.'[10]

As already noted, the importance of *Freedom* and of the theoretical and activist work undertaken by Kropotkin in Britain between 1886 and 1917 has been underestimated to some extent in accounts focusing mainly on the heroic periods of Kropotkin's explorations, imprisonments and activism in Russia and continental Europe. In fact, Kropotkin's British exile was more productive for the anarchist tradition than other periods of his life. Parts, beginnings and anticipations of Kropotkin's classic works on anarchism, such as *Mutual Aid, Modern Science and Anarchism, The Great French Revolution, Fields, Factories and Workshops* and *Ethics*, appeared first as articles in the British 'bourgeois' press, as well as in *The Nineteenth Century*, whereas others, such as *The Conquest of Bread, Anarchism, its Philosophy and Ideal* and *The State: its Historic Role*, likewise appeared as articles in *Freedom* (including some chapters of *Mutual Aid* and *Modern Science and Anarchism*). This linkage between anarchy and geography needs to be fully explained and understood, which is one of the tasks of the present chapter.

According to one of the first historians of *Freedom*, Heiner Becker, the journal was not intended to be partisan, because its founders were primarily committed to opening a dialogue between anarchism and the wider socialist field. While the journal was an important platform to spread Kropotkin's ideas in addition to the role the mainstream press played in this regard, it was mainly intended as an instrument to promote socialist debates in Britain. Many early contributors remained anonymous at the beginning, and a list compiled by Becker exemplifies the pluralistic trend of contributors:

> Among the contributors in the first year were Edward Carpenter, Dr Burns-Gibson, George Bernard Shaw, Havelock Ellis, Sydney Olivier, Saverio Merlino, E. Prowse Reilly, Nannie F. Dryhurst, and Henry Glasse. The non-anarchist contributors were … linked to the Fabian Society and were certainly asked by Charlotte Wilson to contribute, as were others later like Edith Nesbit, or Mrs Podmore who translated Kropotkin's *Conquest of Bread*. [11]

Freedom also acted as a support and propaganda platform for organising periodical meetings on occasions such as May Day, commemorations of the Paris Commune or of the Chicago Martyrs, protests against repression in Spain, in Russia, in Japan, and in other places, which helped gather together migrant anarchists and other anarchist groups. Among the frequent speakers were the Italian anarchist Errico Malatesta (1853–1932) and the French *communarde* Louise Michel (1830–1905). Among *Freedom*'s allies and contributors one finds several activists from the Yiddish-speaking anarchist network in London, who published their own journal *Arbeter Fraynd* (The Worker's Friend) but who collaborated with the *Freedom* group, including Saul Yanovski (1864–1938), Woolf Wess (1861–1946) who later joined *Freedom*, and Rudolf Rocker, a theorist of anarcho-syndicalism, who was Kropotkin's friend and correspondent.[12] Max Nettlau, an Austrian intellectual based in London, former collaborator of Morris's *Commonweal* and correspondent of Kropotkin's, contributed to *Freedom* 'between 1896 and 1913 [redacting] most of the international notes, all the Reviews of the Year, historical and general articles, obituaries, and many reviews'.[13]

Charlotte Mary Wilson (née Martin) (1854–1944), was undoubtedly a major figure in the foundation of anarchism in Britain. The daughter of a doctor, 'she received the best education then available for girls'[14] including studying at Cambridge. Active in several socialist societies, she developed her first interest in anarchism in the 1880s, while siding with William Morris against the choice of leading socialists to participate in political elections, and attending a 'Men and Women's Club, which was organised to arrange frank discussion of sexual problems by the social scientist Karl Pearson',[15] her friend and correspondent. She was also acquainted with significant intellectuals and activists such as Annie Besant, Havelock Ellis, and George Bernard Shaw. Although Kropotkin was a regular contributor and inspiration for *Freedom*, Wilson was effectively the principal editor for the first decade of the journal's publications, authoring most of the editorials and acting as the financial manager and responsible for networking in quest of new contributors. The range of *Freedom*'s contributions and contacts, going beyond a strictly defined anarchist field, shows the importance, for Kropotkin and Wilson, of networking and of disseminating ideas throughout intellectually progressive milieus.

Despite the activism of *Freedom*, early exchanges of letters between Wilson and Kropotkin indicate again how scholarly work and activism were consistently integrated in the creation of Kropotkin's British networks. Wilson's first letter to Kropotkin, surviving at the GARF, shows that she adopted a very cautiously respectful tone as she sought to make contact and to propose a possible collaboration. The deferential style of the letter contrasted with the reality of Wilson's status as a middle-class educated English woman and the condition of Kropotkin, a politically persecuted dissident then imprisoned in France. I quote this document extensively for its revealing nature on some features of their relationship.

It is with much hesitation that I venture to trouble you with this note. I have long considered if, in any practical shape, I could express the respectful sympathy and admiration which I share with so many of my countrymen. It has occurred to me as possible that you may be glad of more help in research, looking out references in the British Museum

Library ... and that I might be useful in this way. I must trust to your kindness to pardon an intrusion, based on a mere supposition of this sort: my reluctance to gesturing upon such a step has been increased by the superficiality of my scientific knowledge and by distrust of my own abilities. But, if it should happen that you would prize such assistance at convenience and if, on testing my powers, you found I could render it ... I should esteem it an honour to work for you. If I can be of no service, please do not trouble to reply.[16]

Future relations between Wilson and Kropotkin would be of a very different kind, but this letter is important, first, because it shows that the two founders of *Freedom* were already in touch before Kropotkin's return to Britain in 1886 and, second, that they agreed on a key point, namely, the need for documentation and for mutual aid in collecting documentation for the anarchist geographer's 'centre of calculation'. This arrangement corresponded with Wilson's personal purposes. According to Walter, her programme for women's emancipation meant that 'she refused to live on her husband's earnings',[17] so some research or editorial work would have been excellent for her to gain an income independently. Furthermore, it seems that Kropotkin's decision to return to London in 1886 was not motivated by mere necessity, but more by the attraction of archives, learned societies, and activists in socialist and anarchist fields for whom Kropotkin already had sympathies.

In an 1889 working paper document without an addressee, otherwise addressed to the *Freedom* editorial committee, Wilson wrote a eulogy of James Blackwell, who had on occasion taken on her editor's duties with 'much business capacity and energy'.[18] Wilson then went on in the same document to express a wish that *Freedom* could reproduce or summarize an article by Kropotkin published in *The Nineteenth Century*, 'to make an article for the July *Freedom*, if Kropotkin has no objection'.[19] This article was intended to be part of a series on the French Revolution to mark the centenary of this event, whose appreciation in anarchist milieus had been heightened and its value confirmed through Kropotkin's works.[20] The proposed article by Kropotkin was finally published as a *Freedom* editorial in August 1889, and others followed. This type of development further confirms the shrewdness of Kropotkin's strategy in engaging within multiple editorial milieus, while working as a full-time anarchist. Content from work for which he was well paid by editors such as Knowles could be redistributed in different print versions and in several other languages to propagate his political viewpoints.

Though Wilson lessened her anarchist commitment towards the end of the century and returned gradually to Fabian socialism, her letters surviving at the GARF show that she never lost contact with Kropotkin. In 1905, she wrote to him from the continent to offer sympathy on the death of Reclus: 'It is a loss to the world but to you and Sofia I am sure it has been a deep grief ... I feel very sad for you'.[21] In the same letter, Wilson also expressed her concern regarding 'the deep feeling and anxieties the state of affairs in Russia must cause you'.[22] Their friendship was sustained beyond work with *Freedom*, as is confirmed by correspondence over subsequent months, touching on personal matters such as illness affecting Wilson, who acknowledged Kropotkin for his 'very kind letter'.[23] Further exchanges show that the two correspondents continued scholarly collaboration as well. Wilson continued to seek out books for Kropotkin's work, and in 1906 she announced a visit to the Kropotkins' house in Bromley.[24] Sources indicate that their correspondence continued until Kropotkin's departure for Russia in 1917, when Wilson wrote that she was 'thinking so much on you and Sophie these last weeks, knowing what the happenings in Russia must mean to you'.[25] Engaged in humanitarian initiatives for war prisoners, Wilson was then still active in many areas and in touch with anarchists during the war years.

Charlotte Wilson was a partisan of women's rights all her life. Like other eminent anarchist women of this generation including Louise Michel, Wilson did not explicitly describe herself as a 'feminist', considering that women's emancipation was inseparable from the emancipation of all humankind. Moreover, these anarchist women generally did not want to be confused with the suffragist movement, because anarchism traditionally rejected the electoral party-political system for both women and men.[26] However, as I show in the next section, *Freedom* was an important platform for hosting writings not only on feminism and women's liberation, but also on sexual emancipation and civil liberties. Moreover, an analysis of *Freedom*'s networks shows that the presence and activism of women in early anarchism was more important and effective than what has been commonly believed.

Anarchism, female activism and women's rights

As Ruth Kinna has observed, some women's activism was visible in the movement known as 'Russian Nihilism' from the 1860s to the 1880s, and 'Kropotkin's claim that women actively entered into nihilism is not controversial'.[27] Among Russian revolutionary exiles, one figure drawing a certain attention, even inspiring the literary fantasy of Oscar Wilde[28] was Vera Zasulich (1849–1919), who had become famous for attempting to kill Fyodor Trevor, the governor of St. Petersburg, responsible for the bloody repression of the Polish uprisings. Perhaps unexpectedly, this notable women attracted the curiosity of some male observers of such activism, even a degree of sympathy, as was the case with Scott Keltie, who asked Kropotkin, then in Switzerland, 'Do you know the bold lady who shot the oppressor in St. Petersburg?'.[29] Kropotkin's answer does not survive, but we know that Zasulich managed to flee to Switzerland, where she became acquainted with Kropotkin, as is apparent from Keltie's enthusiastic and frank comments: 'I am pleased to hear that the heroic Vera Zasulich is safe and secure; as a Scotchman I hope that your companionship is quite platonic'.[30] This puritan remark on the exercise of 'moral caution' regarding this woman unwittingly revealed some of the challenges that politically engaged women posed to the entire way of thinking of European patriarchal society and offers some support for Kinna's claim that 'the nihilists' sexual depravity' was then a widespread commonplace.[31] Furthermore, this correspondence also shows the early personal candour between Keltie and Kropotkin. Kropotkin was still not married and the British editor, within the limits of his own beliefs, was concerned for him as a person and as a friend. A few months later, in October 1878, Kropotkin put an end to Keltie's concerns by marrying Sofia Grigorievna Ananieva-Rabinovich (1856–1938), then a student in Switzerland, who would accompany him throughout the rest of his life. Kropotkin would meet again another nihilist woman from that earlier Russian period, namely, Vera Figner (1852–1942), who was invited to London by the *Freedom* group in 1909, after spending 23 years as a political prisoner in Russia. Released after the 1905 Revolution, Figner gave a lecture introduced and translated by Kropotkin, who, in outlining her biography, insisted on the fact that, prior to becoming 'inevitably' a revolutionary, Figner had struggled to find her place in society as an intellectual woman, and had to go to Zurich, 'the only University then open to women', before returning to Russia and joining the 'going to the people' movement.[32] Kropotkin also recommended Figner to Skilbeck for the publication of an article for *The Nineteenth Century*. [33] Therefore, keeping close links with female activists was important for Kropotkin from his earliest engagements with radical politics.

According to Kinna, in discussing women's emancipation, Kropotkin distinguished between women of different social classes. He considered that the problems of upper-class women claiming equality and individual rights were different, for instance, from the problems of Russian peasant women, who already had an important role in their villages, and primarily lacked not equality with men, but economic emancipation.[34] Shared activism was finally indicated by Kropotkin as an effective means to pursue equality and emancipation, because 'women transformed socialism by their engagement, forging relations with men that were mutually supportive and beneficial because they were established in equal terms'.[35]

The British context in which the *Freedom* group's focus on women's rights was developing was already marked by an interest in this matter. As Selva Varengo has noticed, women's activism and emancipation were not just topics discussed in the Socialist League and in the Fabian Society. Indeed, feminist claims had been already launched 'by Mary Wollstonecraft with the 1792 publication of her *A Vindication of the Rights of Woman*, where the principles of French revolution and English radicalism are applied to women, against … aristocracy, army, church, family and marriage'.[36] The fact that Wollstonecraft had been the wife of William Godwin, considered as a pioneer of anarchism, reinforced the historical link between anarchism and feminism. Therefore, it is not surprising that *Freedom* counted several women among its editors and leaders (Wilson and Dryhurst being the most prominent) and received contributions from the most eminent international anarchist women at that time, such as Emma Goldman, Voltairine De Cleyre, Lizzie Holmes, Lucy Parsons, Louise Michel and many others. Many women appear during the first 30 years of *Freedom* as colleagues and contributors for specific periods, as in the cases of Mabel Besant Hope and Agnes Henry. An Irish woman like Dryhurst, Henry also allowed *Freedom* to operate from her house in Camden from 1893 to 1895.[37] For several years, one 'Miss A.A. Davies', likewise of Irish origin and involved with Irish anticolonial claims, contributed to the journal.[38] Kropotkin's letters to Marsh show that Miss Davies also had editorial responsibilities for *Freedom* at certain unspecified times.[39] In the years immediately preceding World War One, feminist and anti-militarist articles were authored by Lily Gair Wilkinson, and formidable contributions were received from Lilian Wolfe (1875–1974), the wife of then editor Tom Keel, a woman who was defined as having been 'one of the least public but most important figures in the Freedom Press for more than half a century'.[40] Even in later times, the *Freedom* group relied on outstanding female transnational leaders and activists such as Marie-Louise Berneri (1918–1949) in the 1930s and 1940s.[41]

A survey of *Freedom*'s collections shows that several articles addressed matters on female emancipation throughout the period under consideration here. One of the most famous articles, 'Socialism and Sex', was published in April 1887. Anonymous, but probably written by Wilson, it was a review of the book *Socialism and Sex*, published under the acronym 'K.P.' and effectively authored by Karl Pearson (1857–1936), then a mathematics professor at University College London as well as a friend and correspondent of Wilson's.[42] The article's author first blamed the 'unnatural' way in which the institution of marriage regulated relations between the sexes, and then focused on the underestimation of sex and the work of reproduction in the principal schools of socialism, through presenting these ideas as critical concepts for women's liberation, given that 'economic independence … is essential to the moral dignity of each man and woman in a free society. But our present form of sexual relationship is an effectual bar to the attainment of this economic independence by women.'[43] Since the duty of rearing children had been defined as unpaid and unacknowledged work in capitalist society, home duties had become degrading tasks. The author concluded that, with collective organisation, only a limited number of people would be needed to prepare food for an entire community, freeing up significant working and free time for women who were otherwise tied to doing such work: 'Once both men and women gain economic independence, their sexual relations could be freer, and marriage would be no longer regarded as a profession for women'.[44] This idea of sexual freedom obviously scandalised members of Victorian society. However, Pearson's arguments drew on the fact that, in daily life, everyone was aware of the frequency of clandestine relations and street prostitution, and it was this situation that Pearson deemed as truly 'immoral'. The review concluded by considering that Pearson's book was representative of anarchist thinking on the matter and described it as 'the finest declaration which has

appeared in English of Anarchist belief with regard to the difficult and deli-
cate question of which it treats'.[45] It appears likely that the article was written
with an awareness of details circulating a few years earlier in French on the
'free union' entered into by Élisée Reclus's daughters, Magali and Jeannie,
with two young men, without any authorisation from a mayor or a priest,
with only a ceremony attended by friends and relatives, which had excited
heated debates in the French press.[46]

The topic of domestic work as women's slavery was often discussed in
subsequent numbers of *Freedom*. Kropotkin participated in this debate with
an article titled 'Domestic Slavery', translated from French and published by
Freedom in July 1891. One author who appears to have read the article very
cursorily, Susan Hinely, alleges 'Domestic Slavery' is an expression of some
'misogyny', which aims to 'idealize ... the patriarchal home'[47] and views
domestic work an exclusively feminine duty. This may have resulted from a
misunderstanding, or a decontextualized reading of some sentences, because
the aim of Kropotkin's article was precisely to argue against such views.
Other authors such as Selva Varengo had already highlighted a point which
distinguished the *Freedom* group from most unions and socialist groups,
namely, that the anarchists 'did not support the exclusion of women from
pubic workplaces'.[48] In his 1891 paper, Kropotkin principally argued that 'a
Society regenerated by Revolution ... will put an end to domestic slavery; the
last and most tenacious form of servitude, perhaps because it is the most
ancient'.[49] In support of North American women's emancipation movements,
Kropotkin's critique of the exploitative character of women's work could not
be clearer.

> Servant or wife, it is still upon the woman that the man counts to rid him
> of the burden of household work. But the woman also demands, at last,
> her part of the emancipation of humanity. She no longer wishes to
> remain the family beast of burden ... to remain the household cook,
> laundress and housemaid.[50]

It is true that Kropotkin did not explicitly mention the possibility of hus-
bands doing domestic work within households, but it would have simply been
anachronistic to have advanced this option in 1891, in an article intended to
mobilise support among (male and female) workers who spent all day in
factories.

What Kropotkin proposed was the use of machinery to free domestic
workers from household duties, and to complement the proposals of the
author of the 'Socialism and Sex' review concerning the collective organisation of tasks like preparing food. If the time take up by women in these duties
were freed up, then they would have time to take on paid external work to
achieve financial autonomy, as argued for in previous articles, to do intellectual work or whatever they wished. Kropotkin's concluded with an invective
against capitalist society and women's subordination:

Why! Because women's labour counts for nothing. Because in every family there is at least one slave -the mother- and often three or four servants besides, who are supposed to devote their lives to domestic drudgery. Because even those who dream of the enfranchisement of the human race exclude women from their dreams of liberty, and think it almost unworthy of their lofty masculine dignity to think of those-household matters, which they have shunted upon the shoulders of the woman, the drudge of Humanity. To emancipate the woman is not merely to throw open to her the doors of the University, is not to admit her to the Bar or to Parliament. ... To emancipate the woman is to free her from the degrading toil of kitchen and washhouse; to organise the up-bringing of children in such a way as to free the mother and give her leisure. That will be done, for already it is beginning. But let us recognise that a Revolution intoxicated with the most beautiful words about Free-dom, Equality, and Solidarity, whilst it maintains the slavery of the hearth, will not be a Revolution. It will yet remain for the half of Humanity still in domestic slavery, to revolt against the other half.[51]

The topic of domestic slavery was also addressed in public debates. At a Dublin Socialist Union debate, in September 1891, local activist John O'Gorman 'read Kropotkin's article on "Domestic Slavery" from *Freedom*. A very lively debate ensued, several of the speakers (one of whom named Brownrigg is brother of the Roman Catholic Bishop of Ossory, entirely approves the existing domestic institution) fell afoul of our comrade's opi-nions. Fitzpatrick, King and Toomey upheld the Socialist view'.[52] There-fore, there is some evidence that Kropotkin's views on domestic slavery had an effect within public debates and sometimes scandalised the upholders of patriarchy.

Likewise, demands and strikes by working women were suitably supported by the *Freedom* group, as in the case of the 1888 'Match Girls' Strike', a movement entirely women-led against a 'shameful system of oppression [which] was exposed by our comrade Annie Besant in her weekly paper the *Link*'.[53] Besant (1847–1933) was a famous socialist and feminist, later becoming a theosophist, who was in touch with the *Freedom* group and with scholars such as Patrick Geddes. What is important to note is that *Freedom* gave its unconditional support to the strike. The anonymous author of one article (probably Wilson) concluded:

So wretched are the conditions of the masses of working men and women to-day that these scraps and shavings of a compromise with justice look like a great victory. It is but rarely that so much is gained by a strike. The news that the company will declare a dividend of 15 per cent ... instead of 24 per cent, fills us with triumph! ... The strike of the match girls may be forgotten by society, their sufferings may continue, but the moral effect of their action remains.[54]

The role of women's activism within social struggles was a frequent theme in the columns of *Freedom*. In April 1888, one article celebrated the women of the Paris Commune and their contribution in that fight,[55] while, based on reports from Italy, the journal's editors deplored the conditions of girls working in rice fields,[56] the *mondine*, women who played an important role within the history of workers' movements in Italy. In 1897, Kropotkin prepared a lecture on 'The Women's movement in Russia', which was finally delivered by his wife Sofia because Kropotkin became ill, and which focussed especially on the differences between women of different social classes. As mentioned previously, according to the Kropotkins, rural women in Russia already enjoyed a relative autonomy in relation to men, which meant the movement for emancipation there was mainly promoted by women from the wealthier classes.[57]

A central topic in this debate was controversy over marriage. In an 1888 article, written by Wilson, the *Freedom* group reaffirmed the anarchist view that regarded marriage as an institution which basically served to maintain women in a subordinate position. Significantly, the issue of man–woman relationships was considered an intrinsic aspect of the socialist vision, which had to challenge what the author described as 'married or unmarried prostitution as an alternative profession to productive labour'.[58] The viewpoint that saw the man–woman union established by 'our marriage laws'[59] as a form of prostitution and as a legally sanctioned expression of women's dependency had been already suggested by Louise Michel and other activists. In April 1889, a *Freedom* note complained about:

> a raid upon the unhappy prostitutes of the West End. The police may run in any woman in the streets whom they choose to imagine guilty of the crime of solicitation. It is known to every honest person who has reflected thereupon that so long as increasing crowds of women can only get wages on which it is impossible to live and increasing crowds of over fed idlers loaf about seeking the gratification of lust, so long must the buying and selling of human beings continue and police interference only magnifies the evil ... Here are the police again 'suppressing vice' in the interest of the hypocritical respectability which dare not grapple with its causes.[60]

Therefore, in the columns of *Freedom*, expressions of derision against bourgeois moralism, 'respectability' and sex censorship were not unusual. As I explain in the next chapter, similar responses arose among authors such as Henry Salt, Havelock Ellis and Edward Carpenter.

In 1892, a debate with the perhaps questionable title 'Women under Socialism' was organized by the Fabian Society. The two speakers involved, Mrs Grenfell and Ms Ritchie, insisted that economic dependency was the primary cause of women's subjection. In two consecutive notes in response, commentators in *Freedom* agreed with this view in principle but criticised the idea that the state could resolve these problems by paying wages for house work and regulating maternity. State regulation of maternity was considered a despotic measure by anarchists and especially by followers of Reclus and Kropotkin, who were noted adversaries of Malthusianism. Instead, the author of the first note, Agnes Henry, proposed a need for different social conditions to foster the rights of mothers: 'In a society where there were no specially privileged individuals, individualism might not be incompatible with economic independence, or healthy maternity'.[61] Concerning household wages, Henry agreed with the principle that:

> All useful work that contributes to the well-being and happiness of mankind is in a sense either directly or indirectly productive. The mother who brings children into the world is providing society with future workers, and the woman who cooks a dinner is as much a producer in the scientific sense as the gardener who grows the vegetables, or the butcher who provides the meat. Therefore, whatsoever useful work a woman may do is as much deserving of reward as the useful work of men of whatever kind it may be. Whether a woman's work is in the home, in the old-fashioned established way … or outside the home, she is a bread-winner, in the sense of being a contributor to the needs of social life, as truly as any man, and consequently the only point of real difference in her position in a Socialist society would be that man would recognise her right, equally with his own, to claim a just share of those things needful to her full, healthy and happy existence.[62]

The author of the second note, the well-known Italian anarchist Francesco Saverio Merlino (1856–1930) then exiled in London, criticised the idea of paying wages for housework, claiming that it would relegate women to life within domestic walls for a much longer period.

> Instead of furthering, as it is meant to do, the emancipation of women, it would simply check it for many centuries to come. Woman's emancipation has to be attained by … freeing herself from the chains of custom and prejudice, by claiming and asserting her right to live and love freely, and refusing to sell herself for the whole duration of her life to a man's will. Instead of this, we are told that women must become much more than they are now domestic servants; nay, it was actually said in the discussion which followed the reading of the paper, that the man is to be considered as her employer, her master, and the relations between husband and wife should be regulated as those between master and wage-earner, by the law of offer and demand, by the market value of the labour … A thousand times better the present state of things, with all its injustices and crimes.[63]

In both cases, it is worth noting the strong emphasis on ethical and emotional human fulfilment involving happiness, love, justice and health rather than on merely productive relations.

In August 1895, an article signed by 'F.S. Paul' pleaded for sexual liberty, considering Christianity and its sexual prohibitions as responsible for women's subordination. After arguing that 'sexual independence [is the] inalienable right of every human being',[64] the right of women to choose and change their partners and to enjoy the pleasures of love was publicly and explicitly stated, claims that it would be fair to conclude were likely to have been considered 'scandalous' in Victorian England, as further evidenced in material which I quote in the next chapter. Again on sexual matters, *Freedom* joined the campaign against the 'Contagious Diseases Act' promoted by Josephine Butler (1828–1906) and other feminists, describing it as: 'State tyranny adding its gross injustice to the suffering, the degradation, the contumely which social hypocrisy throws on our outcast class'.[65] This Act established that, in certain districts, all (lower-class) women considered as suspected of prostitution could be arbitrarily arrested by the police and forcedly enclosed in hospitals and medicalised. Such a law has also been described as leading to 'medical rape', considering the intrusive nature of the instruments then used for gynaecologic examinations.[66] In 1898, 'Lizzie Moore' (probably Lizzie Holmes) wrote an article on 'Woman's Freedom', insisting on the incompatibility between women's social and economic dependency and the right of all individuals to express freely their feelings and desires.[67] The strongest advocate of 'free love' and the inclusion of topics involving sexuality within anarchist debates was the Russian anarchist (mainly based in the USA) Emma Goldman (1869–1940), who was likewise a contributor to *Freedom* and an acquaintance of Kropotkin.[68]

In 1897, another anarchist woman based in the USA, Voltairine De Cleyre (1866–1912), published a series titled 'American Notes on Freedom', and delivered a series of public addresses in the UK, with the support of the *Freedom* group.[69] In the same period, following a wave of repression in Spain, another important figure for the history of anarchism and of women's emancipation, Teresa Claramunt (1862–1931), arrived in London with a group of Spanish refugees and started collaboration with *Freedom*, publishing a satirical article on the stereotype of 'tenderness' traditionally considered as a feminine virtue.[70] In 1901, a report titled 'The Communist Anarchists and Woman' from the Congress of the Group of International Revolutionary Socialist Students held in Paris was published in *Freedom*. Basically, this report claimed validity for all the social claims of feminist movements, such as economic equality, autonomy of the individual and sexual freedom, due to the links between women's emancipation and the overall social question, such that 'feminism' was substantially considered as a stance already comprised within the label of 'communist anarchism'.[71] In the following years, similar topics were periodically raised in the journal, such as free love,[72] domestic slavery[73] and demands for greater freedom to work for women,[74] as well as the introduction of the term 'womankind'.[75] Well-known anarchist women continued contributing to *Freedom*, such as Lucy Parsons (1853–1942), the widow of one of the Chicago Martyrs.[76] In 1910, *Freedom* published an anthropological study by Élie Reclus titled 'Woman the Creator of Civilisation'.[77] In 1913, a new female contributor who had recently moved from Glasgow to London, Lily Gair Wilkinson (...–1957), published a series of articles on 'Woman's Freedom'[78] in support of similar positions to those of her predecessors, with a special emphasis on criticising the suffragist position, which anarchist women considered as a mistaken means to attain equality. Ferguson considers this series departed somewhat from the generally didactic tone of the journal toward a genre that she calls 'social sketch'[79] mixing propaganda and literary work including a prefiguration of possible anarchist social spaces. These developments indicate the originality and influence of women's work within *Freedom*.

Finally, it is important to highlight a socialist woman who was a friend and correspondent of both Kropotkin and the Recluses. Anne Cobden-Sanderson (1853–1926) was the wife of T.J. Cobden-Sanderson (mentioned earlier), and took part with her husband in the Arts and Crafts movement inspired by William Morris. Standard biographies of Kropotkin by Woodcock, Miller and others only mention Kropotkin's relation with the husband, while primary sources show that Anne corresponded personally with Kropotkin and with Élisée and Paul Reclus. Although Anne was a socialist and suffragette, adopting more moderate positions than the women writing for *Freedom*, she was in touch with Kropotkin for a number of political activities such as the 'Right to Work Committee'[80] and various business concerns related to the 'Morris foundation [which Anne] renews at Hammersmith'.[81] Anne's letters to the Kropotkins, surviving in Moscow, show their lengthy acquaintance, reciprocal visits[82] and shared activities of political socialising, including for example a night at the Socialist Supper Club to hear a speech by Walter Crane.[83] Anne also corresponded extensively with Sofia Kropotkin on private and political matters, and was among the limited number of Western correspondents who continued writing even after Pyotr's death.[84] In a letter without a clear date, but likely originating during the South-African War years, Anne expressed the same unsympathetic views as Kropotkin on the ambiguities of Fabian socialists before the war, and criticised the behaviour of G.B. Shaw, whom Kropotkin also disliked due to his support for the 1899–1902 South African War, as I discuss in the following sections. Anne even claimed that the line of thinking adopted by the Fabian movement would bring it to an end, although she continued to hope in the realisation of a socialist unity where 'anarchists and socialists work in their different ways for a great anti-capitalist and international movement, and to look upon Imperialism as the common [enemy]'.[85] The presence of these diverse kinds of disputants, as both personal and intellectual acquaintances, helps to explain how the anarchists tended to have better relations with a broader range of other socialist groups than in other European countries, at least until the beginning of the twentieth century.

Similarly, Anne Cobden-Sanderson was a friend and a supporter of the Recluses. When Paul Reclus had to seek refuge in Britain, 'Kropotkin recommended Paul Reclus to [the Cobden-Sandersons], where he found a very safe shelter'.[86] During his trip to London in July 1895, Élisée Reclus wrote to his sister from the Cobden-Sanderson country house.[87] Anne would continue her correspondence with Élisée and Paul in the following years. Some letters of hers to Paul Reclus, identified by his pseudonym George Guyou or G.G., written after he left the Cobden-Sanderson house and settled in Edinburgh, survive in the Patrick Geddes Papers at the National Library of Scotland. In these letters, she discussed matters such as a visit to Brussels where she met the elder Reclus, the project of the 'Paris Globe'[88] and her visits to Edinburgh.[89] Finally, Anne authored a chapter for a multi-authored volume, edited by Joseph Ishill in 1927, with recollections on the Reclus brothers. Significantly, she stated that

> It was through Kropotkin (and what better introduction could we have) that we first knew our friend Paul Reclus in England, by whom we later came to know his uncle and his father Élisée and Élie Reclus. [Élisée] came to us in London anxious to find support for the lately established New University at Brussels, but the idea which such a University repre-sented ... Élisée Reclus' sympathy extended to the animal world, and he was and remained a convinced and practicing vegetarian. Élie Reclus and his wife we knew in Brussels, where they were living in 1895. They remain in my memory as a simple and noble couple, who had suffered much, but remained steadfast to their faith which nothing could change or weaken. Their home was simple like themselves ... At once I felt drawn to him by the great sympathy he felt for the coloured races, and his faith in the future, also for his belief in the equality of men and women.[90]

These recollections provide further evidence that the unconventional positions of early anarchist geographers in matters such as gender equality, race issues and animal rights, while not always perceived by recent commentators, were clear to their contemporaries.

For 'subject races' and for Ireland: Nannie Dryhurst and the others

A strenuous anti-colonialist

A little-known woman, but one who was a pillar of the early *Freedom* group, Hannah Ann Robinson (1856–1930), 'was born in June 1856 and spent her youth at 7 Rathmines Road, Dublin'.[91] Still in her youth, she decided to change her name to Nannie Florence and worked as a governess in the west of Ireland and then in London, where she came to know Alfred Robert (Roy) Dryhurst, a British Museum employee and Fabian socialist, whom she mar-ried in 1884. The couple had two daughters, Norah (who became a suffragist) and Sylvia, later known as a writer under the name of Sylvia Lynd (1888–1952). A friend of Wilson, 'with whom they were busy with both anarchists and Fabians',[92] at the time of the foundation of *Freedom*, Nannie Dryhurst is believed to have already distanced herself from the Fabians to become 'an anarchist communist and atheist'.[93] According to her own recollections, she was brought to anarchism directly by Kropotkin, to whom she had been introduced by Wilson at a reception organised by William Morris in the offices of the *Commonweal* journal in 1886:

> Never shall I forget the impression that first sight of Kropotkin made upon me ... The perfect courtesy, always a feature of Kropotkin's manner, first claimed attention, it made you feel that you were in his eyes an individual worthy of his attention; then the clear and penetrating blue eyes, the massive head and flowing patriarchal beard, the squarely erect figure ... were all impressed on me as we exchanged greetings — in French, I suppose. ... Not long afterwards I was told that Kropotkin was interested in starting a monthly paper to be called *Freedom* for the propagation of Anarchist-Communist ideas. I was asked to co-operate — but how? Up to then my ideals had never gone beyond 'Freedom in Ireland' and the history of her seven centuries of struggle against English domination. 'The very thing we want' said Kropotkin. 'We should know what is going on in the extreme west of Europe'. So, I undertook to take notes for the movement, which at that time was agitating Ireland. ... It was long afterwards that Kropotkin told me how much he enjoyed those Irish notes which I had penned with the gay irresponsibility of youth.[94]

From that moment, Dryhurst became a committed collaborator at *Freedom*, undertaking a full range of editorial tasks as well as being an active propagandist, and standing in as editor when Wilson was too busy. In the 1890s, 'her daughter, Sylvia, would see her laying out the paper on the dining-room table at 11 Downshire Hill'.[95] She also spoke to a meeting at the Berner Street Club (East End) on behalf of the Women's Educational Union in 1891, which also reveals her commitment to achieving women's rights. In his recollections, her neighbour George Sturt witnessed a festive and very cosmopolitan event, attended by Kropotkin, that she organised in Hampstead: 'A sort international social evening where national songs were sung by French and Germans, Russians, Italians, Spaniards and a Cuban refugee: at which Mrs. Dryhurst acted as stage manager'.[96] Dryhurst was active in social and educational initiatives in her neighbourhood. One anecdote concerned a discussion where she shocked some of the attendees 'by announcing herself an atheist'.[97] In Victorian England, radical and unconventional statements sounded even more shocking when they came from women than from men.

Dryhurst was especially passionate about Irish issues, joining the Hampstead branch of the Gaelic League[98] and remaining a radical supporter of Irish independence after finishing as a direct contributor to *Freedom*. Another of Nannie Dryhurst's characteristics was her multilingualism, being proficient in French, German and Irish, while her archives suggest that she could read Italian, and later in life she tried learning Georgian. This proficiency allowed her to work as a translator, an obviously useful skill for the anarchist cause. In 1896, she corresponded with Nettlau on possible translations for *Freedom*;[99] in 1909, she translated Kropotkin's *The Great French Revolution* into English from its French book version (even though Kropotkin's first works on this subject had been published in English by *The Nineteenth Century*). Her correspondence with Nettlau also demonstrates her prominent role among the 'Old *Freedom* staff'; on the occasion of some personal problems which might have slowed down her activity, Dryhurst urged Nettlau, recently arrived in the group, to take responsibility along with others: 'you younger members must take up that … soon',[100] which is revealing for her attitude to leadership.

Among her activities, her teaching at the International School in Fitzroy Square played a significant role, immersing her in anarchist education, and especially in welcoming the children of migrants. The school had been opened in 1891 by the most famous anarchist woman in London and probably in the world at that time, Louise Michel (1830–1905).[101] Michel was a survivor of the 1871 Paris Commune and of deportation to New Caledonia, and lived in London in the 1890s. According to Bantman, life for French anarchist exiles living in London in those years was especially hard due to language problems and social inclusion difficulties. In this context, Dryhurst was one of Michel's rare friends in London and a pillar of the school together with other *Freedom* activists such as Wilson and Henry.[102] Dryhurst's command of French was decisive for her role in liaising with the community of French anarchists: some letters surviving at the International Institute of Social History (IISH) show that she usually wrote letters in French. In 1903, she outlined to Michel, then in France, the pedagogical methods of a school she had recently visited.[103]

Some typewritten biographical notes surviving in the Dryhurst's family papers at the National Library of Ireland in Dublin highlight her acquaintance with Varlaam Cherkezishvili, known as Tcherkesoff (1846–1925), and his Dutch wife Frida. A Georgian aristocrat (like Kropotkin, he was born a prince) who committed to anarchism along with Reclus and Kropotkin, and who later fought for the independence of his country,[104] Tcherkesoff was likewise an intellectual, a friend of Reclus and Grave and a contributor to *Temps Nouveaux*. In London, Tcherkesoff contributed to *Freedom* with several scholarly series such as 'Pages of Socialist History' later published in book form,[105] and various writings on the history of the International and on Russian imperialism. One Tcherkesoff 'particularity' was his implacable opposition to Marxism, which indeed characterised most of the anarchist communists, who were very averse to being confused with authoritarian communists. One example of Tcherkesoff's writing in this regard was his caustic critique of Marx and Engel's *Communist Manifesto*, which Tcherkesoff considered a work plagiarized from a pamphlet by French socialist Victor Considérant.[106] Very committed to the cause of nationalities oppressed within the Russian Empire, Varlaam Tcherkesoff remained Dryhurst's close collaborator and life-long friend, during a period when her primary concerns involved the anti-colonialist struggles of the Irish and of 'small nations'. In

1930, Frida Tcherkesoff assisted Dryhurst in her last days and wrote her obituary for *Freedom*. [107] In terms of personal friendship, the archival notes also highlight relations between Dryhurst and the Kropotkins: 'Apart from the Tcherkesoffs, who arrived in London in 1892 originally, her most important political friendship was with Peter Kropotkin. He and his wife and daughter were frequently at 11 Downshire Hill, and they wrote affectionate letters as well as some politically enlightening ones'.[108] The sources I consulted indicate that Dryhurst's concerns for Ireland and for 'small nations' were substantially shared by Kropotkin.

In the 1890s (or late 1880s), Dryhurst began a long affair with journalist Henry Nevinson (1856–1941). Nevinson's diary documents this moving story of love and passion in the context of late Victorian society, which obliged the two of them, both married (at a time when the right to divorce was severely restricted), to conceal their relationship. The story is recounted in some wonderful pages of a biography of Nevinson by Angela John. According to John, heated political discussions, especially on the Irish question, formed a part of the clandestine couple's relationship, which led to a rupture after 20 years.[109] However, Nevinson seemed to have been strongly influenced by the approach of his lover on political issues. A member of the Social Democratic Federation, Nevinson became a friend of the anarchists and corresponded with Kropotkin.[110] He attended 'anarchist meetings' and collaborated anonymously with *Freedom* and with Louise Michel's school, always under the guidance of Dryhurst, whom John describes as 'a leading light in London's anarchist circles, one of several remarkable women connected with *Freedom*'.[111] According to John, Nevinson also associated with Edward Carpenter and tried to take French classes from Louise Michel, seemingly with poor results. Nevinson's own recollections state that he was 'intimate with the Anarchists during these years', having 'a friendship lasting for many years with two remarkable people: Louise Michel and Peter Kropotkin'.[112] These recollections also provide an example of how British sympathisers of anarchism were impressed and sometimes intimidated by the figure of Kropotkin, and secretly shared jokes about his English pronunciation, as on one occasion when they wondered why one should strive for 'the abolition of all *low*'.[113] So deeply did these major anarchist figures influence Nevinson's life that almost 40 years later, when *Freedom* resumed publications following an interrupted period, he publicly acknowledged the return of 'my friend and ally of nearly forty years ago ... No paper is more needed in these days when tyrants, oligarchs and bureaucrats are so triumphant'.[114] While Dryhurst was described in several recollections as an especially charming and 'young-looking' woman, it was more her strong personality which allowed her to exercise a charismatic role among her closest acquaintances.

In autumn 1906, with the help of Varlaam Tcherkesoff, Dryhurst travelled to Georgia to document Russian repression after the 1905 uprisings. On her return, she presented a petition 'signed by thousands of Georgians'[115] to the Peace Conference held at The Hague in 1907. At her house in Hampstead,

Dryhurst established a headquarters for efforts to organise a petition on behalf of Georgian women denouncing massacres and rapes by Russian soldiers, which was signed by more than 3,000 English women.[116] According to the previously mentioned biographical notes from the Dryhurst family papers, it was at that time that Dryhurst and Nevinson 'formed the Committee for the Defence of Nationalities and Subject Peoples in order to be able to work happily together and for a good cause'.[117] A 'Subject Peoples [or Races] International Committee'[118] was indeed founded in 1907, and at its first executive committee meeting, Nevinson acted as chairman and Nannie Dryhurst as secretary. This committee's statutes stated that it was to act only as a coordinating organisation, and not interfere in the political decision-making of other national sections, which initially included the Aborigines Protection Society, the British and Foreign Anti-Slavery Society, the Egyptian Committee, the Friends of Russian Freedom, the Georgian Relief Committee, the National Council of Ireland and other pacifist societies. Among the rights asserted by the International Congress of Subject Races held at The Hague in August 1907, it was declared 'that the claim of every subject race of distinctive nationality to the management of its own local affairs shall be recognized by the Dominant Power'.[119] While this proposed right might appear as rather moderate, the assertion that 'the rights of subject races and population in times of civil war, rebellion of other disturbance shall be identical with the rights of belligerents'[120] clearly implied substantial support for the right to anti-colonial insurrection. These shared activities of Nevinson and Dryhurst can be considered as anti-colonialist, with reference to both 'internal' and 'external' colonialism.

Dryhurst organised a conference on 'Nationalities and Subject Races', held at Caxton Hall in Westminster in June 1910, whose proceedings were published in a volume, which in turn can be considered, in association with a book edited by Jean Grave in Paris in 1903 and prefaced by Élisée Reclus, *Patriotisme et colonisation*, as involving the first collective expressions of political anti-colonialism in both countries, respectively.[121] In her preface to the English volume, Dryhurst complemented the thrust of Reclus's and Kropotkin's anti-colonial arguments questioning the tenets of 'civilisation'. Dryhurst demanded consideration for 'those obscure peoples, for whose welfare so few care, and upon whom for long centuries the most wrong have been inflicted by men claiming to belong to higher races ... probing the canker of modern civilisation',[122] and was enraged at the repression exercised by colonial authorities. The Irish anarchist especially denounced 'such condemnations as Farid Bey's [which] will be impossible within the jurisdiction of any people claiming to be free'.[123] Egyptian nationalist Mohammed Farid Bey (1868–1919) had been imprisoned in Cairo for publication offences, and at the 1910 conference, he had delivered one of the more memorable speeches, protesting against the progressive transformation of the British protectorate into an exercise of British absolute power within the internal affairs of his country.[124] During a session dedicated to the Indian delegation, according to attendees, 'Bipin Chandra Pall, made a speech of such passionate venom against the British that the people in the front row had to cover their knees with their mackintoshes'.[125] Among the participants it is worth noting the name of Swiss human rights activists René de Claparède (1862–1930), the founder of the Swiss League for the Defence of Indigenous Peoples and author, concurrently with Roger Casement (1864–1916), of the first public denunciations of Belgian atrocities in the Congo.[126] Other colonised or subaltern nations represented included Finland, Poland, Georgia (by Tcherkesoff), Persia, Morocco (by Nevinson), and of course Ireland. A specific study is needed of this event, but it is already possible to claim, first, that the initiative for it to proceed was informed by the conference slogan reproduced in the dedication of the published book: 'If we fail to denounce the crime we become participators'. Second, Dryhurst played a principal role in these committees, also attending the Egyptian National Congress held in Brussels in 1910, where she was received as 'the Irish delegate, and made the proposal that a sub-committee should be set up representing India, Egypt and Ireland, the subject peoples most oppressed'.[127] In 1906, *Freedom* published a note on Nevinson's correspondence and anti-slavery activities in Africa, where he was working for a North American journal:

> H. Nevinson, the traveller and journalist, who through Harper's Magazine of New York has been exposing the hideous cruelty of the Portuguese in their West African colonies—cruelty on a par with the Congo atrocities—reports that since his articles were published an American factory owner has stated that he will not manufacture another grain of cocoa imported from the colonies in question.[128]

It is clear that these activities were not remote from the concerns of the *Freedom* group. In a short biography of Dryhurst by Nick Heath, it is suggested that Dryhurst had distanced herself from the anarchists at that time. However, newly available sources suggest the opposite. First, one of her letters to Nettlau dated November 1907 reveals that Dryhurst was sending him Georgian-related material, while, in the same document, she mentions a recent visit to Kropotkin to discuss these matters, and Tcherkesoff's activism on the same subject.[129] Throughout these years, her correspondence with Kropotkin was prolific. While the British Library[130] and the Dryhurst family papers[131] hold only a few letters from Kropotkin to Dryhurst from 1893 to 1915, a folder filled with letters written by Dryhurst to Kropotkin from 1906 to 1917 survives in Moscow. Significantly, these letters are concentrated in the period when it has been claimed she had left the anarchist movement, and show that, even though she no longer acted as *Freedom*'s editor while she busied herself with the *Subject Races* initiative, she never lost contact with Kropotkin, Tcherkesoff and other radicals. Dryhurst and Kropotkin exchanged an enormous number of letters in the period 1908 to 1909 in relation to Dryhurst's translation of *The Great French Revolution*. The tone of their correspondence reveals Dryhurst's deference and respect towards Kropotkin, but also a great confidence in discussing family and personal (including confidential) issues in addition to business matters.

In 1906, Dryhurst discussed with Kropotkin ways for circulating information about the repression in Russia that 'Knowles refused to publish',[132] and in the following years she liaised with the publisher Heinemann in relation to the process for publishing *The Great French Revolution*, [133] inviting Kropotkin several times to visit her in Hampstead to discuss these topics.[134] On publication of Kropotkin's work, Dryhurst expressed her pleasure at Kropotkin's achievement with a letter written on stamped paper of the Gaelic League of London.[135] In 1908, some of Dryhurst's observations in her letters to Kropotkin reveal an increasing concern with the struggle for the liberation of Ireland, expressed in her prediction of 'some move in Ireland soon. The *Sinn Fein* [activists] are quickly preparing the people for it'.[136] Some disagreement on the principle of nationalism seemed already to be emerging, as one might surmise in a note from Dryhurst telling Kropotkin that: 'You have no conception of the poverty among Irish nationalists. The Unionists invest every penny.'[137] The Irish cause steadily became Dryhurst's main motivation, to the extent that she broke with Nevinson in autumn 1912. According to John, their final quarrel was due to a split in the Irish Women's Social and Political Union, while 'disagreement over Ulster precipitated the parting. Nannie had told Henry to leave the subject alone because he could not understand it.'[138] Considering the forthcoming events, she was probably right.

In those years, during her frequent trips to Dublin, Dryhurst was involved with initiatives of the feminist association *Inghinidhe Na hÉireann* (Daughters of Ireland) and was in touch with Countess Markiewicz and probably with Maud Gonne. Meanwhile,

> it was at a Gaelic League meeting that [Nannie's] daughter Sylvia met Robert Lynd, a nationalist rebel from Belfast, whom she was later to marry. He in turn introduced the Dryhursts to Roger Casement, and one of the last letters he received in Limburg (where he was unsuccessfully trying to recruit for Ireland Irishmen in the British army who had been taken prisoner by the Germans) came from NFD.[139]

Dryhurst was a close friend of at least two of the 16 'martyrs' of the 1916 Easter Rising and the following repression, namely, Thomas MacDonagh (1878–1916), who had come to know her on the occasion of her Subject Races lectures in Dublin in 1908,[140] and Casement, and after the Rising, 'her energies were devoted to the campaign for [Casement's] reprieve'.[141] Beyond his activities in favour of the Congo's indigenous peoples, Casement was one of the most active militants for Irish independence.[142] In May 1915, Dryhurst wrote to him very cordially.[143] In that same year, Kropotkin contacted Dryhurst to know her opinion on the war, and she responded rather evasively, wishing the end of 'all imperial robberies'.[144] Clearly it would appear in her case, from a perspective concerned more with Irish self-determination, that the struggle of the 'Western democracies' against the central empires was much less defensible than for Kropotkin. In July 1916, Dryhurst wrote an emotional letter to Kropotkin requesting urgent support to save Casement. Kropotkin's response is not available in the archives, but some embarrassment may perhaps be supposed because, for the first time in their correspondence, Dryhurst used harsh nationalist language, writing that, from London, she looked back 'at the list of illustrious rebels who have obtained welcome and refuge here and contraste[d] it with the English fire in Dublin and the atrocious treatment of Roger Casement. I begin to [wonder] whether English people have any sense of justice at all'.[145] It is unlikely that Kropotkin, welcomed by the 'English people' for more than 40 years, could have shared this way of viewing the issue. Nevertheless, Dryhurst's last surviving message from April 1917 is written in a spirit of warm friendship, celebrating 'this great wave of freedom that has passed on Russia'[146] and wishing Kropotkin and his family the best of luck for their return to Russia after their long exile. On this occasion, Dryhurst was writing from Dublin where she was apparently still very energetic and rejoiced in 'plenty of intellectual companionship'. As I show in the following section, Irish independence was likewise one of the main overall concerns of *Freedom*, mainly due to Nannie Dryhurst.

Freedom *for Ireland*

In the first decades of *Freedom*'s publications, interest in the Irish question was high and covering news and debates from Ireland were particularly important activities. *Freedom* started its publications in October 1886, when public interest in the 'first Home Rule crisis' which had resulted in the fall of the Liberal government led by Gladstone, was temporarily declining in public debates. Nevertheless, the first years of the journal are particularly rich with notes and information related to Ireland, which were included in almost all the journal's issues. Even though these articles and notes were mostly anonymous or signed with pseudonyms, it is possible to suppose that, at least in the 1880s and 1890s, they were principally due to Dryhurst, and, to a lesser extent, to some Irish correspondents whom it is not always easy to identify, although meeting notifications in Dublin for activist propaganda did give the names of some socialists and by extension, given their intellectual proximity then, of some anarchists. Kropotkin's interest and direct involvement in these discussion is shown not only by his frequent mention of Ireland in his work and in Dryhurst's recollections, but also in the letters he sent to other members of the group, for example to Alfred Marsh, to call attention to the Irish question: 'Could you say just a word about this book by Morrison-Davidson.[147] Something about Home Rule all bound being welcome and consequently all breaking up of turn huge machines, the States.'[148] Later, Kropotkin also expressed some sympathy for the 'Celtic revival'. A letter he sent in 1904 to one of the principal figures in this movement, William Butler Yeats (1865–1939), survives at the National Library of Ireland, and not only expresses Kropotkin's interest in one of Yeats's plays, but also reveals an earlier acquaintance between the Irish poet and Kropotkin's family, as Kropotkin sent regards 'from all three of us'.[149] Kropotkin and Yeats were apparently introduced 'at Morris's house',[150] while scholarship in the history of literature shows that Kropotkin influenced Yeats's political ideas, as he considered sincere revolutionaries like Kropotkin as an inspiration 'for our national propaganda'.[151] The other great Irish national poet and novelist, James Joyce, was also a reader of Kropotkin and Bakunin,[152] as well as of Reclus and Metchnikoff.[153] It can be assumed that, in Ireland, Reclus's and Kropotkin's works had that double readership which characterised their circulation in most countries, namely, a readership of progressive intellectuals interested in geography for various reasons, and a readership of workers reaching out for socialist and anarchist information and inspiration.

In 1886–1887, the failure of the 'parliamentary way' to Irish Home Rule gave further support to anarchist anti-parliamentarianism. Among the first observations on Ireland published in *Freedom*, an article titled 'Parliamentarianism and Revolt' analysed the political defeat of the followers of Charles S. Parnell (1843–1891), leader of the Irish Parliamentary Party, arguing that a 'political' revolution would not have resolved the social question. Although class struggle was central for anarchists, it would be mistaken to conclude that they neglected national and colonial issues, as the colonial nature of British domination in Ireland was clearly examined in *Freedom*, as revealed in statements such as: 'The Irish people have been conquered by the English. Like all subject races they have been shamefully wronged and oppressed by their foreign masters. For hundreds of years they have been struggling to free themselves from this yoke.'[154] Furthermore, what the anarchists proposed for resolving Ireland's economic problems was to continue the social struggle in the Irish countryside, where peasants were inspired 'by the brave spirit of insubordination which hundreds of years of tyranny have been unable to quell. Ireland is leading the van of the universal land war, and she is leading it by the only means … by which the masses of people in any land and any age have successfully withstood robbery and oppression: direct, personal, open resistance.'[155] This Irish struggle paralleled the struggle of English workers whose interests, for the *Freedom* group, had likewise to be supported by direct action rather than through the ballet box. What the elements of a possible anarchist formula for Irish freedom could involve needed then to be elucidated.

A series of 12 short articles published anonymously in each number from October 1886 to February 1888, but written by Dryhurst, was titled 'Law and Order in Ireland. How they were introduced'. It started with a history of the 'conquest of Ireland by the King of England', replete with mentions of local resistance and reminding readers of the responsibilities of the Catholic Church which had allowed the first occupation of the island in the twelfth century, also defined as the 'Pope's gift'.[156] This history paid special attention to the introduction of land property structures, including some biting remarks on the origins of 'the parliamentary farce',[157] later called 'imposture'.[158] Indeed, Ireland was often evoked as a case study to demonstrate the value of popular insubordination and unruly attitudes. In August 1887, an article on 'Spontaneity' opened with a quote which was supposed to be familiar to the reader: '"If there's a government, then I'm agin it". We are emphatically of the opinion of that oft quoted Irishman.'[159] The 'Law and Order' series was then supplemented with periodical observations on the Irish situation. In 1887, an article titled 'Home Rule and after. Impression of an English anarchist in Ireland' provided commentary on landlordism and poverty, and a rather appreciative discussion of the views of Michael Davitt (1846–1906), considered as more radical than the supporters of Parnell in matters of economic struggle.

Overall, it appears that the national cause and the social question were seen as interlinked: 'At present as a nation, Ireland stands on the eve of the realisation of her hopes. The dreams which her poets have dreamed and the visions which her younger sons have always seen, are to be dreams and visions no longer. She is to be a nation, a "United Ireland", governing herself and working on her own salvation'.[160] Some observations made concerned the alleged 'natural poverty' of Ireland, an idea which was very widespread, including within the Irish working classes, and was considered as contributing to an alleged general resignation in the face of a miserable destiny. According to *Freedom*'s editors, what would lead the popular classes to revolt were the mistakes of parliamentary politics. The lack of results from the efforts of politicians would have persuaded 'the workers that the movement which shall give them class freedom, or economic independence, must emanate from themselves'.[161] In the same number of *Freedom*, a note on 'Coercion and Revolt in Ireland' praised local resistance against various Coercion Acts of the previous decades, arguing that 'the revolt of the Irish people against foreign dictatorship and land monopoly grows daily more effective'.[162] In the following issue, mention is made of the struggles of Irish nationalist William O'Brien (1852–1928)[163] and a comparison made between the anarchist martyrs of Chicago, involving five workers executed in Illinois in November 1886 after a politically motivated trial,[164] and another event, the 'November Martyrs', involving three Irishmen hanged on 23 November 1867 in front of Salford Gaol, near Manchester.[165] From an anarchist standpoint, these comparisons clearly indicated that the message British anarchists wished to send to Irish people entailed: 'our struggle is your struggle'. An effective synthesis of the anarchist position on Home Rule was contained in an 1889 article 'Ireland for the workers', which claimed that: 'As Anarchists, we are Home rulers in every sense of the word.'[166] This meant that an anarchist view of Home Rule implied not only the removal of colonialism, but also of capitalism and the state.

In monthly notes and reports on Ireland published in 1888 under the heading 'The Struggle for Freedom', there are condemnations of the workhouse system 'crowded with the victims of evictions' and concerns raised regarding the growing extent of Irish emigration. Given this situation, anarchist hopes were considered to rest in local agency and defiance: 'The axe has been laid at the roots of landlordism. The people have combined to save themselves from extermination, and are doing so, not only without the operation of government but despite it.'[167] In these notes, as in other cases of early anarchist and republican anti-colonialism at that time,[168] there are scathing observations on the pretentions of the dominant powers to bring 'civilisation', which meant in practice that alleged 'savages' were 'hunted down, fined, imprisoned, bludgeoned, and bayoneted'.[169] Reports on evictions and abuses resulting from local landlords were often published in *Freedom*. This suggests that Dryhurst and her friends could count on correspondents based in different Irish counties. An Irish 'Song of Rebellion' was published in November 1888.[170] In the following years, correspondence from local socialist clubs increasingly appeared. In December 1889, a Paris Commune commemoration was held at the Dublin Socialist Club in Marlborough Street, a debate on 'Anarchism versus Social Democracy' took place at the Progressive Club,[171] and similar meetings were periodically reported over the following years. In 1893, the creation of an anarchist group in Belfast was announced.[172]

In 1891, *Freedom* launched further critiques on Parnell and the institutional Home rulers,[173] including a withering note on 'Home Rule and Rome Rule', which ironically re-cast the slogan of Protestants fearing the power of the Catholic Church should Irish autonomy be advanced. According to the anarchists, the Catholic Church was doing all it could to substantiate this objection with its reactionary and obscurantist behaviour, including 'scurrilous attacks on everyone who did not join the clerical throng'.[174] A disquieting note in June 1891 offered some clarity as to why the Catholic Church was a problem for socialist and anarchist propaganda in Ireland at that time: 'The upshot of this idea may be the formation of an Anti-clerical party, but Irishmen have such a terror of what John Mitchell used to call "their pauper souls" that they cannot be relied upon to oppose the priests for any considerable time.'[175] While this note might be considered as undermining the Irish progressive movement compared to 'stronger' ones such as British or French progressive movements, it is worth noting that abundant literature[176] has highlighted how the Catholic Church was and remains a substantial issue in relation to autonomy, modernisation and respect for civil rights in Ireland, a problem which is also apparent in the recent debates on the movement 'Repeal the Eight'. Everything that opposed clerical tendencies in Ireland was praised in *Freedom*, for example, the new 'Independent Party, who have struck a good note on their resistance to priestly influence'[177] or the association of 'The Irish National Schoolteachers [who] have at least dared to unite together to protect themselves against clerical despotism'.[178]

Over many years, material on Ireland was continually presented in the pages of *Freedom*, ranging from the sombre commemoration of the centenary of the 1798 uprising,[179] to the publication of Irish songs in May 1900.[180] In the 1890s, according to Bantman, there was some common ground between anarchists and Fenians in the milieus of London-based exiles, despite the improbability of shared aims, as was also stated in police sources, such as being involved in a project for 'blowing up Westminster'.[181] In 1906, the emergence of *Sinn Fein* was welcomed by *Freedom* with some interest, hoping that it 'may eventually develop in a Socialist and anti-militarist direction'.[182] An empathic tribute to Irish independence movements was also paid by anti-colonialist militant Guy Aldred in an article against parliamentarianism in 1908,[183] in a period when *Freedom* was mobilised in favour of Scottish anarchist John McAra (1870–1915), imprisoned in Belfast during a propaganda tour. In several articles published over the 1908–1909 period, a correspondent who signed as 'The Irish Rebel' engaged in a debate with Guy Aldred on the question of the struggle for parliamentary control.[184] These engagements provide evidence of how often the terms 'Ireland' and 'rebellion' were associated in the anarchist press of that time.

Finally, *Freedom* covered the events of the 1916 Easter Rising, although London anarchists were not able to provide more concrete support to this movement due to the state of war and their limited numbers. In the same period, the group had lost the support of Kropotkin, who had promoted a very controversial Manifesto of the Sixteen in solidarity with the cause of the 'Western Democracies' in the ongoing world war, while the *Freedom* group retained its radical anti-militarist positions.[185] In 1914 and 1915, reviews and reports of James Connolly's works[186] appeared in *Freedom*, praising the future Irish martyr and reader of Kropotkin[187] for his analysis of the war and of the responsibilities of the working class in relation to it. According to one reviewer, Connolly had recognised 'that the signal for war ought also to have been the signal for rebellion'.[188] While repression in Ireland had already been condemned by *Freedom* in 1915,[189] several articles were published in 1916 claiming the right of Irish people to revolt and to celebrate the Easter martyrs.[190] It is not possible to delve more deeply into this subject because it would go beyond the scope of this book. However, in respect of both Nannie Dryhurst and on *Freedom*'s commitment to Ireland, further studies are recommended.

This section has shown an ambivalence within anarchism in matters of nationalism and national liberation, as already discussed in works addressing cases of 'oppressed' nationalities in the geographical works of Reclus and Kropotkin.[191] On the one hand, when the nation does not coincide with the state, it is not considered incompatible with anarchism. On the other hand, the national cause makes sense for anarchism only if it opens spaces for social transformation, not limited to a simple substitution of a statist and capitalist power for another one. These critical and varyingly nuanced considerations, however, did not prevent anarchists from supporting all the major anti-colonialist struggles in Europe and outside Europe in the age of high imperialism, as I discuss further in the following sections.

Kropotkin and Alfred Marsh: between activism and scholarship

Alfred Marsh (1858–1914) is another little-known activist who nevertheless played a very important role in the *Freedom* group as its principal editor from 1895 (when Wilson was no longer able to manage the journal) until his death in 1914. According to Nick Heath, Marsh being an obscure worker did not prevent him from gaining prestige which impacted his most famous peers: 'He was clear sighted about Kropotkin's faults and ... was one of the few people who could make Kropotkin understand what was not possible'.[192] In Marsh's obituary, Tom Keell, his successor as *Freedom*'s editor, wrote that the journal's 'existence today is almost solely due to [Marsh's] courage and his faith in Anarchism. His pen and his purse were always at its service, and on several occasions his last half-sovereign ensured the publication of the paper, especially during the Jingo reaction of 1899–1902, when the movement was at its lowest ebb'.[193] This raises another significant aspect of anarchist histories: the success and duration of groups and journals were often decided more by the voluntary and hard-headed commitment to very motivated individuals and small groups, rather than by 'structural' conditions. These efforts were often localised, such as in towns or neighbourhoods where some favourable environment could exist and establish lasting roots, which introduces the historical geographies of anarchism, a concept recently launched in geographical literature.[194]

The correspondence between Kropotkin and Marsh is one of the few cases in which a conspicuous number of letters from both sides are available in public archives, in Moscow in the case of Marsh's letters, and in Amsterdam in the case of Kropotkin's. However, more than 100 of Kropotkin's letters survive at the IISH, which is much more than Marsh's 30 in Moscow. Thus, also, in this case, the researcher encounters a certain asymmetry of the source. Marsh took over *Freedom*'s editorship in 1895: though Kropotkin was less involved in the journal's business after 1900, also due to ill health, the correspondence between the two men shows that the Anarchist Prince never gave up his action in the group, which often occurred by way of Marsh as his informal 'representative'. Kropotkin's first letters in 1895 were replete with compliments and encouragements for Marsh's work, and confirmed the strict relation between *Freedom* and its French 'twin', *Les Temps Nouveaux*, inaugurated then in Paris by Jean Grave (1854–1939) following the series *Le Révolté/La Révolte* and was especially cherished by Kropotkin.[195] A few years later, Marsh was considering Kropotkin's suggestion for enriching *Freedom* with a 'literary supplement'[196] on the model of what Jean Grave was doing in Paris with *Temps Nouveaux*.[197] On the occasion of his 1897 trip to North America, Kropotkin requested Marsh to 'tell the comrades a few farewell words' and to 'take the best care of *Freedom*', committing to provide the journal with 'interesting materials, either in the form of leaders or correspondence'.[198] In 1902, Kropotkin proposed new ways for doing *Freedom* editorial meetings to allow editors assessing the materials in the future, due to a certain sloppiness in the reviewing process that the Anarchist Prince lamented.[199] This means that, through Marsh, Kropotkin remained involved in reading and assessing the materials to be published in *Freedom* for all, or at least for most of the period from 1886 to 1914, when he finally quit the group. Therefore, over this period, *Freedom*'s positions were unlikely to be a far cry from Kropotkin's ideas, unless when they were presented as public controversy.

This collaboration included informal consultations between the two men in internal matters and issues of the *Freedom* group, as Marsh had often to 'represent' Kropotkin during the long periods of absence for health reasons that the Russian geographer took in milder localities such as Rapallo, Locarno, or Bordighera, very frequently after 1905. In 1912, Marsh took Kropotkin's help on some ill-defined editorial controversies with two members of the 'second generation' of collaborators, Nettlau and Tcherkesoff, hoping that the prestige of the Anarchist Prince would strengthen his own position.[200] The intimacy between Kropotkin and Marsh is also confirmed by their confidence in discussing matters of family and health. On an amusing note, the Anarchist Prince attributed his periodical states of illness to a disease he acquired while in French prisons. 'Fever each time returns. The dear Clairvaux malaria!'[201]

What is also worth considering is that the relationship between Marsh and Kropotkin was reinforced by shared editorial business not directly related to *Freedom*. Marsh helped Kropotkin redact a glossary for one of his works,[202] and enjoyed collaboration with *Freedom*'s editor on his *Nineteenth Century* affairs. In 1896, for instance, Marsh was requested to forward materials from the *Geographical Journal* for the redaction of Knowles' periodical.[203] Scholarly contributions were still appreciated as militant devices by activists, as Marsh explained to Kropotkin, matching Wilson's arguments mentioned above: 'Your work on the *French Revolution* ought to have an immensely wide circulation, I believe it will, for the subject is first of all undying interest to all the peoples, and the fact that you have thrown new light on a neglected side of it will lead as many to a renewed study of an all-absorbing subject.'[204] In this vein, Marsh collaborated with Kropotkin for the book edition of *Modern Science and Anarchism* by *Freedom* Press,[205] constantly arguing for the need of Kropotkin's works 'appearing in English'.[206]

During Kropotkin's sojourns abroad, Marsh kept him informed on the situation of the political struggle in Britain. At the end of 1909, the editor of *Freedom* was not very confident in revolutionary novelties. 'I don't think we have the men to rouse the country, although it must be confessed that it is first on such issues that England has started her previous revolutions.'[207] This exposes how history was considered as an inspiration for future revolutionary hopes. A 1910 letter from Marsh was the occasion to assess the numbers of *Freedom*'s dissemination: though the British anarchist mentioned only 2,500 copies printed for the incoming issue (they would become 3,000 to 3,500 in the following years),[208] he related his efforts for rendering more accessible the journal and printing of special propaganda material for the May Day. 'The contents also will be more varied, and very readable for the great public. As there will be probably 40,000 or 50,000 people in the park ... we have decided to print 20,000 of a new leaflet "What is Anarchism" for free distribution. I think you will like it'.[209] As for the spread of ideas in the intellectual world, Marsh was still optimistic, and he greatly praised his new co-editor: '[Thomas Henry] Keell, works splendidly ... Again, the call for anarchist literature and speakers is remarkable. I do not remember anything like it in the past 25 years.'[210] This was also one of the criticisms *Freedom* attracted by its detractors, including in socialist and anarchist fields: the charge of being a middle-class, intellectual organ not easily understandable by a not better defined 'people'.[211] If this was surely a limitation of Kropotkin's strategy, already highlighted by other anarchist communists less committed to theoretical work like Malatesta,[212] these documents demonstrate that intellectual anarchists did not spare any effort to reach comprehensive and popular dissemination with all the means available to them.

Some samples of the scientific and evolutionist topics addressed in *Freedom* in the years of Marsh's editorship confirm the importance of geography, scholarship and Kropotkin's influence, especially in the numerous references which were paid to evolutionism and mutual aid. Prestigious scientists such as Wallace and Spencer were often quoted, and extracts of their papers sometimes published in the journal. Wallace was unsurprisingly mentioned in association with Darwin and other scientists as one of the exponents of the new scientific methods challenging religion and dogma.[213] Spencer was the object of a series by Kropotkin on co-operation in reply to *Principles of Sociology*, a work which, according to Kropotkin, 'deserve[d] our full attention'.[214] Kropotkin also devoted a series of articles to Spencer's thinking after Spencer's death.[215] The differences between their respective philosophies have been widely discussed by Kropotkin's vast literature quoted above: what is worth noting here is that Spencer was considered by the *Freedom* group as an author whose writings could serve progressive political propaganda. It was the case with an excerpt from Spencer's *Man versus the State*, against state socialism, published by *Freedom* in June 1903.[216] Darwinism was the object of a paper series signed Jehan Le Vagre (an obvious pseudonym for Jean Grave), an author who was all but a theoretician, which testifies to the deep penetration of Darwinism associated with the theory of mutual aid in popular anarchist propaganda.[217] In May 1891, a sarcastic note criticised the famous Italian positivist and anthropologists Cesare Lombroso for his book *The Criminal Man*, where the author indicated anarchism as a criminal psychopathology identifiable by the physiognomic traits of the concerned individuals. Therefore, for early anarchist scholars, modern science could provide services to the cause, but it had never to be taken as a new dogma or new religion.

Geography was often mentioned in the journal, and some significant notes were also devoted to exploration; despite the frequent critiques of the overseas behaviour of explorers, functionaries, militaries, and priests by the anarchist geographers, some explorers had relations with anarchism or enjoyed sympathies among anarchists. It was the case with Arctic explorer Fridtjof

Nansen (1861–1930), whom Kropotkin met at Geddes's house in Edinburgh in 1886.[218] As I explain in the next section, Nansen's writings were used by *Freedom* as an example of empathy towards native peoples (eventually the Inuit) facing conquest and extermination. In 1897, Nansen was criticised by some mainstream journals and found perhaps unexpected support in *Freedom*'s editors, through a note probably written by Marsh:

> Now that it has come to light that Nansen is to receive close upon £15,000 for a description of his journey to the silent ice-regions and a work upon his observations there, we find the well-informed journalist of the Mammon-worshipping character is glibly writing of the consolations of North-Pole expeditions, as though Nansen undertook this brave and perilous task as a purely financial speculation. We wonder whether any of these gentlemen would have sufficient moral and physical courage to undertake such a journey even as a financial speculation. Let us recollect Nansen is an explorer and human scientist first, and that financial results in his case were purely accidental, or rather circumstantial … The innumerable dangers encountered in his journey, the swim for life, his never-ending battle with the elements, his readiness to risk life itself, his simple and honest description of the same in the *Daily Chronicle*, are all sufficient answers to the worthless insinuations of that corrupt journalism which now attempts to cover his grand work with the nauseous cloak of commercialism.[219]

After decades of postcolonial literature, reading anarchists vehemently defending an explorer against the bourgeois press might look amazing, but it is understandable in the light of the existence of early anti-colonial and anti-racist milieus which included geographers and travellers, as I discuss in the following section. Ironically, Nansen would indirectly reward the anarchists for their sympathy because, when he became a diplomat and launched the 'Nansen passports' for political refugees, several anarchists including Emma Goldman took advantage of them in the 1920s and 1930s.[220]

In any case, activism continued to complement Kropotkin's scholarly interests in his British years, and intensified significantly in the years between the two Russian revolutions, from 1905 to 1917. In that period, other exiles from the Russian Empire, like Tcherkesoff, helped Kropotkin in reconstructing networks with Russia and in circulating information to western revolutionary movements through the columns of *Freedom*. In May 1905, Kropotkin wrote to Marsh that he received: 'Good news from Tch[erkesoff]. He writes that whole groups of Russian Social Democrats are going to join anarchism.'[221] Although Kropotkin's optimism would be deceived by the failure of this revolutionary process and by the following repression, anarchist historiography confirms that 1905 was the year which saw the organisation of the first anarchist groups bearing this name operating on Russian soil, while in the former decades Russian anarchists had essentially operated in exile.[222]

The insufficiency of theoretical work was one of the drives which moved Kropotkin to come back to Russia in 1917 and which could already be perceived in 1905 by remarks such as 'English life is so dull, so dull! This is why I feel nothing that I might write about with interest.'[223] With Russia in ebullition, the Anarchist Prince was no longer excited by English activist debates and preferred to publish in *Freedom* on history, geography or international agendas.

In 1911, Marsh sent to Kropotkin reassuring news on the health of Errico Malatesta, still in London.[224] Malatesta would become the adversary of Kropotkin in the 1914–1916 controversy about the support given by the Anarchist Prince to Western democracies against the Central Empires. Even though the reasons of Kropotkin's controversial choice have been widely debated, and recently well explained by Ruth Kinna,[225] it is worth noting that, during WWI, the great majority of British anarchists 'took up a strongly anti-militarist position',[226] and this eventually entailed Kropotkin's de facto resignation from the *Freedom* group. However, the correspondence between Kropotkin and Marsh addressed antimilitarist topics until the eve of the war. Although Marsh died in 1914 without seeing the outcomes of this discussion, his last letters reflect this growing concern. In 1913, he wrote to Kropotkin that 'the rise of the anti-militarist spirit should be important to check the warmongers',[227] adding that he was translating an old antimilitarist brochure by Kropotkin himself, *La Guerre*, [228] accordingly the series 'Modern Wars and Capitalism' was published by *Freedom* in 1913, a year when Kropotkin still had occasion to 'call in town' and meet with *Freedom* people.[229] This also indicated that the war controversy was grounded on tactical rather than on 'doctrinal' bases, with antimilitarism being a concept always claimed in principle by Kropotkin. In any case, Marsh could not join the final controversy. In an obituary published in November 1914, Kropotkin defined Marsh as 'a brother'.[230]

Decolonizing socialism (and geography): The Black Man's Burden

Reclus and Freedom *against the Empire*

Recent literature has highlighted the strong anti-colonial drive of early anarchism and anarchist geographers,[231] and the contributions which transnational anarchism gave to the dialogue between radicals of European and non-European cultures[232] and to 'anti-colonial imagination'.[233] A survey of the papers published by *Freedom* between 1886 and 1916 (and possibly later) confirms these findings, revealing how anarchists used geography to conduct early criticisms of internal and external colonialism, racism and Euro-centrism. Scholars and activists belonging to these circuits were key players in transnational networks of decolonial solidarity and paid attention to what today is called the rights of minorities. Here, my argument is that if for a long time, anarchism was not considered among anti-colonialist movements, this was mainly due to the unfamiliarity of most non-anarchist (and some anarchist) scholars with anarchism. Anarchism rarely displayed the label of 'anti-colonialism', considering national liberation as a part of a wide social struggle; nevertheless, anarchists gave constant support and solidarity to anti-colonial causes in a way that was generally earlier and more radical than in the case of all other socialist schools of European origin. Notably, in this case, geography mattered.

As the first step of this survey, it is worth considering the pioneering contributions of the Recluses, especially Élisée, who discussed anti-colonial topics in his correspondences with London anarchist Henry Seymour. The 1885 letter that I quote below, published in Reclus's correspondences in 1911, was also translated for *The Anarchist* and included in *Le Révolté*. For Reclus (as for Kropotkin), the question of colonialism was first associated with militarism and patriotism. Writing to a British comrade like Seymour, Reclus first felt the need to clarify his internationalist views by stating: 'They taught me to hate you with the pretext of patriotism, as your nationalist patriots learned to hate the stranger ... But our country is greater than what our masters wanted.'[234] Coming to an assessment of the behaviour of European nations in an era of high colonialism (1885 was the year of the Berlin conference which split up Africa) he presented his arguments with his usual sarcasm:

> To judge the moral value of conquering nations, it is enough to look at the work of European states discussing how to portion the world. They look like ravens gathering around a corpse and taking one piece each. It is what in common language they call 'The triumph of civilisation upon barbarity'. Raid and massacre are both brave exploits which cannot but render proud the fellow citizens of thefts and murderers. We learn that thousands of men were put to the sword, that villages were burnt, that horse's feet opened human chests.[235]

Being French (though exiled), Reclus considered that he had first the duty of addressing his homeland's misfits, that is, French colonialism in Northern and Western Africa and South-Eastern Asia.

> The nation harshly paid for these crimes. Likewise, she paid for her triumphs in Algeria, when brilliant officers, accustomed to massacre Arabs and Kabyle, came back to Paris to execute other "savages" [the Communards] sweeping popular neighbourhoods with their ordnance, as they did with the poor Arabs' villages. France will also pay for Tonkin and Formosa. The flows of history will bring the punishment for the fault committed.[236]

Reclus's reference to the employ of colonial troops in the massacre of the Parisians at the end of the 1871 Paris Commune evoked one of the shocks which spread the first elements of anti-colonial sensibility in the French worker's movements. Anti-colonial ideas also circulated after the Communards' experience of deportation in New Caledonia from 1872 to 1880, when Louise Michel sided with the anti-colonial revolts of natives, even scandalising most of her fellow Communard exiles, still conditioned by racist stereotypes about the Kanaks.[237]

Reclus's eyes were constantly turned beyond the Channel. As he wrote to Seymour: 'But is France the only coupable in Europe? England has her odds which will fall back on her, consuming her like a cancer. Her violent annexation of Ireland will imply her punishment day by day. And your immense colonial empire will not make the happiness of your patriots.'[238] Whilst Ireland's case can be considered as 'internal colonialism',[239] the chapter on the British Empire that Reclus wrote in *L'Homme et la Terre* analysed all the facets of British imperialism. Reclus first fumed at jingoism and the 1899–1902 South-African war. If national fanaticism was a problem everywhere, Reclus observed that, following those events, 'it is probably in England that patriotism took its most deliriant and hard form'.[240] Reclus was well aware of the responsibilities of men like Cecil Rhodes (1853–1902) in African wars and colonial massacres. Today, Rhodes's name is the object of the anti-colonial campaign 'Rhodes must fall' to remove the monuments celebrating people responsible for colonial crimes in Britain and the colonies.[241] 'Cecil Rhodes, who gained this way hundreds of millions ... put as the absolute starting point of his conduct: "I establish that we are the first people in the world"'.[242] Unlike some anarchists and socialists who sympathised with the Boers during the 1899–1902 Anglo-Boer War, considering them as exponents of a free republic which challenged the biggest Empire in the world,[243] Reclus was well aware of the crimes committed by all Europeans, British, German, French and Dutch, against the native populations of Austral Africa. Thus, he considered this conflict as only 'the occasion of deep disagreement in the great Church of imperialism',[244] that is an inter-imperial conflict. According to Reclus, the idea 'that all the world will be one day devolved as a prey'[245] characterised the British but could relate to every imperial power (Figure 4.1).

Characteristic of Reclus's anti-colonialism was his geopolitical analyses which sometimes anticipated elements of future decolonisation, stressing the lack of demographic weight, and occasionally the disorganisation, of what is now called 'invaded colonies', which rendered the oppressors vulnerable when the oppressed gained consciousness of their potential force (Figure 4.2). For Reclus, it was the case in Austral Africa, where 'intelligent Bantu ... witnessed the weakness of their masters'.[246] There, the need of keeping a numerous army for controlling the Boers, the Afrikander and for 'massacring the hated Kaffirs' implied that 'these colonies menace to become a new Ireland'.[247] Likewise, the British domination of the Indian peninsula was considered by Reclus as the premise for violent clashes. 'The enormous Indian empire was submitted by violence, and it is still subdued by material force and all the complementary apparatus of canons, rifles, tribunals and prisons. [Now] this oppression is carried on with expert ability, with a great knowledge of people, smartly opposing nationalities, playing as the referee of all disputes. [This] must result in nefarious consequences for both masters and servants'.[248] The final aspect of anti-colonial resistance which Reclus mentioned,

Figure 4.1 Anti-colonial caricature from Reclus's last book
HT, vol. VI, 77

anticipating later works by James C. Scott,[249] was the persistence of indigenous communities refusing connections with the so-called 'civilised' world, eventually in the hinterlands of Indochina and Indonesia, where 'so called savage tribes occupy more than the half of the territory [and] resist morally to the conquest'.[250] For Reclus, the idea that modernity disorganised communitarian customs of different societies could be compared with the social transformation which enclosures and industrial capitalism entailed in modern Europe, presenting imperialism as opposed to traditions of communitarian mutual aid.

These pages of Reclus's are a pertinent introduction to the anti-colonial notes and papers published in *Freedom*, approximately in the same period. In 1887, the first *Freedom* paper explicitly devoted to a colonial issue concerned a 'minor' colonial power, Italy, but contained significant comparisons with the major ones. The Italian attempts to subjugate Abyssinia were read as the nth attempt of capitalism's expansion through the spoliation of new peoples.

L'Irlande est bien le vautour qui
ronge le flanc du Prométhée britannique,

Figure 4.2 L'Angleterre et son cortège [England, and her court]
It is worth noting the caption: 'Ireland is the vulture gnawing at the side of the British Prometheus'.
HT, vol. VI, 1

> For some time, Italy has rung with the cant so familiar to English ears, about carrying the blessing of civilization to savage nations. Exploring parties were dispatched. Finally, General Pozzolini was sent 'on mission' to King John. The old, old story we English know so well; explorers, missionaries, traders, land grabbing, exploitation, and then armies and artillery to enforce the submission of the 'barbarians' to the tyranny of the whites who rob and enslave them.[251]

It is worth introducing a terminological warning: at that time, terms such as 'civilised', 'primitives' and 'barbarians', were generally used by anarchists without any moral judgement of superiority or inferiority. In most cases, they simply indicated the distinction between industrialised and non-industrialised regions; in this case, they were often used sarcastically, to denounce the alleged 'civilisation' the colonisers brought to natives.[252] After lamenting the weakness of anti-colonial mobilisations in Italy until that moment, the *Freedom* paper concluded by considering an international social revolution as the best antidote against colonial crimes. 'A social revolution which shall free the working classes of Europe and America is the only hope of peace and well-being for those races which have as yet escaped the blessing of exploitation, brandy, and vice.'[253]

Other topics characterising Reclus's geography were his anti-racism and his struggle against slavery in North and South America.[254] In 1888, the last big nation allowing slavery, Brazil, abolished it, and *Freedom* first praised the fact that 'a million slaves obtained their liberty in Brazil'.[255] However, the anarchist journal tried to 'deconstruct', as one would say today, the official narratives praising the paternalistic philanthropy of Brazilian empress Isabel as the reason for this decision. Based on French translations of a Portuguese socialist journal, the editors of *Freedom* insisted on the numerous riots which preceded the formal liberation of slaves, insisting on what is called today subaltern agency as the only way to 'strike the blow'. Recent scholarship on Brazilian quilombos, communities of fugitive slaves in a state of perennial revolt, and generalised slaves' insubordination confirms these views.[256] Finally, *Freedom*'s editors discussed the back side of the medal of this liberation, eventually the dramatic conditions of waged workers in Brazilian plantations, arguing that: 'The Brazilian slaves have forced their way out of the frying-pan only to find themselves in the fire.'[257] If the historiography of transnational anarchism shows that European migrants appointed in Brazilian *fazendas* in the following years revolted considering themselves to be the new slaves,[258] historiography on Afro-Brazilian movements shows that the abolition of slavery and the proclamation of the Brazilian republic did not resolve the problems of these communities, especially in regions like the Northeast, where revolts continued.[259]

Freedom denounced some situations where European 'civilisation' proved its evils and its contradictions. In March 1889, a paper commemorating the Paris Commune mentioned a recent British colonial massacre in Africa to counter the argument that 'evil' Communards had set fire to Parisian buildings for reasons different than military defence. The argument was set with the usual sarcasm:

> A few days ago, the writer of this article read a telegram in a daily London journal stating that the British troops in the Soudan had set fire to 5,000 huts belonging to the unfortunate natives. Doubtless, this was excusable as it was done by brave British soldiers and professedly in this interest of war. But a great deal is made of the statement that the communists fired and destroyed buildings in Paris.[260]

In a paper on the appropriation of land in the nineteenth century, there were mentions of the colonial drive of capital accumulation: 'Men who had made fortunes in business—especially the businesses of wringing wealth from the wretched natives of India and the colonies, from the slave trade and from usury—returned from abroad with the one object of buying land.'[261] In July

1889, correspondence from South Africa discussed the problems of socialist propaganda, which hardly reached the 'coloured portion of the proletariat', while the white one 'was afflicted by race and colour prejudices'.[262]

In 1891, a correspondence by Lizzie Holmes from Chicago explained the situation of North-American Indians, arguing that:

> No race has been so systematically wronged as have the genuine American people—the original owners of the soil. From the time they ran down to meet the wonderful white men with gifts and sniffing welcome on the Atlantic shores, to the last outrage committed against them on the frontier, their kindness has been met by robbery, their trust with treachery, their rightful demands with murder and extermination.[263]

From the story of early conquest until the segregation and extermination the natives still suffered at that time, this article was a long invective against the crimes of 'civilisation', including an appreciation for 'primitive' institutions.

> As to civilisation, it is to their credit that they refuse to receive it, with all its attendant evils, vices, wrongs, disease, and poverty. They are too brave and free to take upon themselves the bonds that civilisation's poor must bear. An intelligent, free civilisation that recognised individual rights would reach them, for they have with them the possibilities of a great, strong, enlightened and noble people.[264]

This kind of material confirms how unjustified were the claims of some so-called 'post-anarchists', accordingly unfamiliar with the anarchist tradition, who mistakenly defined 'classical anarchism' as part of the colonial project of modernity, as discussed.[265] Most early anarchists appreciated modernity as far as it got rid of former dogmas but also contested its violent aspects.

In the columns of *Freedom*, impunity for colonial crimes was presented as an example of the unevenness of bourgeois justice, as in the case of an 1891 correspondence from Italy relating a trial which had taken place in Massaua, the main port of colonised Eritrea. 'There some officers and government officials were accused of wholesale massacre and plunder of poor natives under the most frightful circumstances. The facts have been ascertained by a Parliamentary Commission. Yet the court has acquitted everyone. Oh Anarchy of Justice! Oh Justice of Anarchy!'[266] The asymmetry of international (and therefore internal) justice was also denounced in correspondence on the opium culture in India, a business on which it was said that: 'England cannot deny her share in the wholesale poisoning of China'.[267] The *Freedom* paper noted that not only selling opium in China but also cultivating it in India had a demoralising effect on local societies, by increasing diseases, mortality, suicides and violence.

> Not content with wrecking the art of India, we must demoralise its people. Truly we are a good, a Christian civilisation! A triumph of

progress, and a glory to the nineteenth century! Yes, we carve one nation with the sword, and poison another wit opium; and breed parasite upon parasite up through all the castes of society, we live upon the labour and degradation of the poor; and to crown infamy build up towers, temples and gorgeous' palaces for scoundrels and 'such like'.[268]

The opium wars in China were likewise condemned by Reclus, who observed that common stereotypes of barbarianism and civilisation were applied to the European invaders, in the case of a Chinese definition of the English militaries imposing this trade as the 'Red-haired barbarians'.[269]

Jingoes and Matabele

The most important anti-colonial campaigns of *Freedom* concerned the British colonial wars in Austral Africa and tried to organise workers' resistance against militarism and jingoism. These campaigns started with an 1893 sarcastic note on the accusations of crimes which colonial powers in Africa were mutually exchanging with one another, which ridiculed the alleged philanthropic reasons which were mobilised to prepare the Matabele war in the region later called Rhodesia.

> Every Briton is of course properly disgusted by the abominable French exploitation of the Siamese, and the shameless invention of imaginary grievances as an excuse for final annexation. But most Britons don't recognise that England is engaged with equal greed and still more hypocrisy in preparing to annex a big slice of South Africa. ... The English are so damnably clever in inventing pretexts. We generally set about stealing land from native people with a plausible pretence of philanthropy. Just now ... the astute English are going to war in Mashona land to protect the poor oppressed Mashonas from that cruel tyrant Lobengula. ... But the fate alike of Lobengula with his conquering Matabele, and of their victims, the Mashonas, is certain eventually to be that of the luckless oysters ... to have made excellent meal.[270]

Indeed, pretexts for humanitarian interventions and 'right wars' were very similar to those employed in modern wars related to what geographers call today 'the colonial present'.[271]

Lobengula Khumalo (1845–1894) was the king of the Matabele, whose massacre and the related responsibilities by Cecil Rhodes and friends were denounced by Reclus,[272] Kropotkin[273] and of course by *Freedom*. In February 1896, an editorial titled 'War' denounced the increasing spreading of a nationalist, colonialist, and militarist mentality, including among proletarians. The imperial endeavour was ironically compared to the ancient Garibaldian cause, one which was surely worthier in anarchist views.

If there were a drop of heroic blood left in the middle-classes of Europe, a Garibaldi would have long since landed, red flag in hand, in Armenia. But no! The admiration of the modern man is for those who arm the Bechuana police ... for seizing gold mines which might raise by so many hundred pounds the South African shares of the West End millionaires. The modern Garibaldian is the English clerk who shoots down the unarmed Matabele 'like dogs' and takes their cattle which will be divided among all of us, or goes to fish military crosses and promotion with the Ashantis. And when a flying squadron is armed in a hurry, and volunteers are mobilised, it is to conquer the territory on the Orinoco, given up in 1841 by Lord Aberdeen, when gold mines on it were a mere myth, but which is claimed now, as a dear property, as soon as gold is discovered on it.[274]

A note titled 'English Atrocities' denounced the 'terrible sufferings and premature death to which workers in white lead factories are subjects'[275] in South Africa; while the author of 'Chivalry and barbarism' fumed again at Cecil Rhodes: 'We have read of the Jameson raid; we know what sort of stuff Rhodes and the Chartered Gang are made of. If there is such a thing as scum in human society, surely these are the scummiest of the scum'.[276] Needless to say that, in the mainstream press, the term 'scum' basically identified the poor.

Despite the difficulties in obtaining first-hand information from reliable sources in Africa at that time, *Freedom* tried to release news not only on colonial crimes but subaltern resistance. The Matabele uprising was defined as a 'revolt of an outraged people against civilised barbarities',[277] arguing that 'these barbarous savages have actually dared to defend their homes and families'.[278] The value of rebellion for all the colonised was also exposed by Reclus in the Belgian socialist journal *Humanité Nouvelle*: 'This hatred of the slave who revolts against us is right, and proves at least that there is still hope of emancipation. It is natural that the Hindus, Egyptians, Kaffirs and Irishmen hate Englishmen; it is natural that Arabians execrate Europeans. That's justice!'.[279]

It is in this context that two excerpts from Fridtjof Nansen's *Eskimo Life* were published by *Freedom* to counter the jingoist rhetoric on the superiority of the 'civilised'. The first contained a sarcastic invective against 'our civilisation and our missions' and their historic role in destroying subjected civilisations.

What has become of the Indians? What of the once haughty Mexicans or the highly gifted Incas of Peru? Where are the aborigines of Tasmania and the native races of Australia? Soon there will not be a single one of them left to raise an accusing voice against the race which has brought them to destruction. And Africa? Yes, it, too, is to be Christianized; we have already begun to plunder it, and if the Negroes are not more tenacious of life than the other races, they will doubtless go the same way when once Christianity comes upon them with all its colours flying. Yet we are in no way deterred, and are ever ready, with high-sounding phrases about bringing to the poor savages the blessings of Christianity and civilization.[280]

The second one, 'An uncivilised view of Civilisation', matched indirectly Kropotkin's thesis of morality as a consequence of natural tendencies towards mutual aid and not as the result of political doctrines or religions. Nansen had this experience among Greenlander 'Eskimos', whom he defended vehemently against despising judgements of Christians and the 'civilised'.[281] Inuit life was also one of the favourite topics of Élie Reclus, an author whom *Freedom* saluted at his death by stating that: 'No man in Europe has known and *understood* primitive man so well as [him]'.[282] The 'primitive communism' which Reclus envisaged in Arctic communities was discussed again by *Freedom* in 1910.[283]

In the period preceding the Boer war, reports on atrocities committed in the colonies of different European countries became increasingly frequent. In 1897, a note titled 'Savages?' denounced the complicity between Christian missions and European businesses, especially in Africa, and raised the problem of reaching the Black masses with socialist propaganda: 'There must be other missionaries who shall go out to these people, and preach the gospel of Solidarity and Fraternity instead of Rum and the Bible'.[284] It is worth noting that the same language was then used about the need of propaganda among white proletarians, so this note did not imply any pretension of ethnic superiority. In the same year, *Freedom* launched one of the earliest denunciations (preceding both Conrad's *Heart of Darkness* and the famous Casement Report) of the Belgian atrocities in the Congo:

> We are living in an age of cruelty. Emperors and ministers un-blushingly order the destruction of races and places to both display their fearlessness and their power. It must be fresh within the memories of everyone, how 'honourable' officers in the Belgian army have shown their worthiness of their posts by valorously cutting off hands and feet from natives in the Congo and committing other unmentionable atrocities in honour of H.R. H. King Leopold ... East African atrocities are not dissimilar to those on the Congo.... Never in the history of ancient or modern civilisation has there been such a league of human bandits, reeking with the blood of murdered people and speedily wearing the world to ruin.[285]

Discussing imperial rivalries between France and Japan on some small islands of the Pacific, *Freedom*'s editors were reminiscent of Denis Diderot's *Supplément au voyage de Bougainville,* considered by Sankar Muthu as a manifesto of 'anti-colonial Enlightenment',[286] and of the systematic destruction of small insular communities by European sailors and missionaries, in questioning again the very concept of civilisation.

> A hundred years ago the natives of Tahiti lived in a similar way, free and happy on their islands in the Pacific, until discovered by the Europeans,

when civilisation (read missionaries, drink, syphilis and general misery) was introduced and the happy race debased and ruined. It reads like a sort of vengeance when to-day a few Europeans, trying to live a happy life on such islands, are forced back under the sway of *civilisation* by a people considered but a few years ago almost savage themselves, but now, as the Japanese are, eager to ruin themselves as quickly as possible by overdoses of civilisation! Where at present civilisation treads it cannot but crush and ruin.[287]

Finally, a sarcastic note titled 'Saving the Soudan' reported the outcomes of the war carried on by Anglo-Egyptian troops there. After crushing Dervishes and Atbara, these 'criminals' were advancing in their 'work of demolition', significantly said to disregard local geographies and cultural differences.

Again, the Soudan is not distinguished from other parts of the globe by its geographical position alone. There are its peoples, their languages and their customs all of which are the chief means whereby we arrive at their distinction from other races. And it is just all this we are destroying and bent on completely annihilating. It is the Europeanisation of its land and people that is aimed at.[288]

At the eve of the Boer conflict, it was a *Freedom* editorial by Élisée Reclus, reproduced in Appendix B, which provided something like an 'official line' of British-based anarchists for protesting against the war, patriotism and colonial crimes, phenomena which they all saw as strictly interconnected.[289]

The 1899–1902 war marked a difficult period for *Freedom* activists, who campaigned against imperialism and militarism in almost complete isolation. If in anarchist milieus there was some sympathy for Boer 'free republics', what was clear from reading *Freedom* was that its editors denounced the responsibilities of all European nations, not only England, in the massacre of natives.

The whole history of the conquest of the Black Continent by all Europeans without exception—British included—is a history of murder, poisoning, extermination, deceit, slave-buying and slave-making—so much so that Africa may be described as a true school of barbarism, under all its possible aspects, for all European nations. It was so, and still is for the English, the French, the Germans, the Italians, the Dutch. So much so that even when workingmen of advanced opinion have gone to settle in Africa many of them have become there as complete brutes as any one of the capitalists, in their relations towards the natives. We will never forget the fury with which Malatesta, on his return from Egypt, told us of the treatment he saw inflicted there by Italian workingmen—more or less Socialist at home—upon Egyptian fellaheen. Leaving aside Invidia cages ... leaving aside such brutes as Stanley or Peters—what did the

English do, as a nation, with the sanction of Parliament, in Matabele Land?[290]

It is worth noting that this kind of observation dispelled dreams about workers' settlements in 'tropical' or 'unpopulated' lands, which characterised several socialistic schools throughout the nineteenth century. Though South African early Whites' settlements were not organised with a state sanction, 'the horrors which British and Dutch settlers have committed upon the Black in the Cape Colony at the beginning of this century were revolting'.[291]

The March–April 1900 issue was almost entirely dedicated to the war. The opening article on 'The British Workers and the War' analysed the economic situation, stressing how shameful it was that: 'The Blacks have been brought by English law into serfdom, and are compelled in Kimberley to work for whatever wages a company chooses to pay them'.[292] The responsibility for this situation was attributed 'to the greediness of the Rothschilds, the De Beers, the Rhodeses, the Chamberlains and other international bloodsuckers of the Lombard streets of London and other European capitals'.[293] The most bitter and problematic aspect of these campaigns, according to the anarchists, was the lack of any massive opposition in the British working class, whose majority did not counter the colonial wars.

> When the open town of Alexandria in Egypt was bombarded by the British ironclads, without even showing the mere pretext of 'resistance', and when the church bells rang all over England to glorify this massacre, where were the English workers to protest against this act of highway-robbery in which France refused to join? When the Matabeles were shot 'like nine-pins', their cattle were taken, and serfdom ... was imposed upon the survivors, what did the British workers say to that? They gave their approval.[294]

According to the *Freedom* group, this was the result of suicide politics carried on by the majority of socialist and union leaders, who supported the South African campaign or did not overtly oppose it. Anarchists' polemics were especially directed against the Fabians and their leaders like George Bernard Shaw, despite their former good relations with him. After internal dissensions, which *Freedom* ironically compared with mental diseases, for the Fabians 'imperialism practically became the test-question [and] resulted in a substantial majority in favour of the Rhodes-cum-Chamberlain policy of crushing the Boers'.[295] Polemically, *Freedom* also republished the excerpt of a text opposed to colonisation which Shaw had published some years earlier.[296] Anarchists' position on this point was summarised by Kropotkin in a letter to Marsh, blaming reformists and Fabian socialists and using harsh tones against Shaw, defined by Kropotkin as 'that ... acrobat'.[297] The Anarchist Prince considered British non-anarchist socialists as generally 'imperialist', arguing that some of them joined ranks with anti-war movements in 1901

only because '[Henry] Hyndman has always been against the policy of spoliation in India and in Egypt'.[298] Other letters state how Kropotkin disliked
Shaw's behaviour, defining him a 'Philistine'.[299] In fact, despite Kropotkin's
biographies often mentioning Shaw as one of the most famous of his relations
in Britain, Shaw's archives only contain one letter sent by Kropotkin in 1903,
basically discussing literary topics.[300] This contrasted with the fact that the
Fabians were traditionally closer to anarchists than Hyndman's parliamentary
socialists, and confirmed the centrality of the anti-colonial issue, on which
anarchists claimed their originality and radicalism concerning the remainder
of the socialist field.

Even in those years, *Freedom*'s criticism did not neglect to document
colonial crimes by nations other than England. In 1900, a long section of
the 'International Notes' was devoted to the misfits which took place in
French colonies, like the massacre in Martinique of a sugarcane planta-
tion's workers who had protested against misery, and the 'massacre of the
Sakalaves in Madagascar'[301] by the staff of General Gallieni. In the same
period, a formerly important colonial empire, the Spanish one, was losing
its last overseas territories. In countries like Cuba, anarchists took part in
anti-colonial movements, and some *Freedom* correspondents from the
Antillean island claimed the success of this cause: 'The revolutionary proletarians after their heroic struggle against Spanish tyranny in favour of
political independence, are now prepared to fight the politicians of Cuba
and the United States in order to secure social emancipation and absolute
liberty.'[302] A similar situation characterised the young republic of the
Philippines, which had just thrown off the yoke of Spanish domination
when it was confronted by another decolonised republic, but with a very
different political weight, the United States of America. *Freedom* praised
the Filipinos as 'brave fighters in a cause forlorn. The living protest of a
race not wholly eaten by corruption'.[303] The arrogance of US imperialism
in the Pacific was compared to the British one in South Africa, stressing
the paradox of the fact the two most powerful nations in the world did
not find anything better to do than to crush small nations which 'only
want to live independently'.[304]

In May 1901, a text by Mark Twain was reproduced under the title of
'Imperialism pilloried',[305] and in the following journal issue an appeal from
French revolutionary syndicalists to British workers, translated into English,
called for international proletarian unity against the dangers of war, including
anti-colonial concerns:

> The world is not limited to our two nations, and never perhaps has war
> made greater ravages than at the present moment. No government is free
> from the reproach of having committed odious acts. Blood is flowing in
> North and South Africa as also in Madagascar, the Philippine islands,
> etc. In China, the allied troops of the West have indulged in the most
> revolting massacres ever recorded in history.[306]

Meanwhile, the *Freedom* campaign to stop the atrocities in South-Africa continued with an appeal 'To the people' published in December 1901, calling to resistance against 'these titled and empowered murderers [who] squandered 250 millions of your money and sacrificed 20,000 of your brothers and sons "for the glory of England" they say! For the benefit of Rhodesian money grubbers; for the introduction of serfdom under the British flag among the natives.'[307]

In June 1902, the end of the war was not saluted as an improvement, as the editors of *Freedom* feared that the British government might try new imperial endeavours. Meanwhile, the outcomes of the war for South African society, and especially for the Black community, were painfully lamented:

> For becoming the only masters of millions of blacks, and after having destroyed their village-communities, taken their land and cattle, and imposed upon them a tax which they are quite unable to pay—to deliver these blacks to the gold mining companies, saying to them: Here are plenty of cattle for you. Make that cattle work for any wages you like, or no wages if you prefer! They are your serfs—slavery is antiquated, you know, but serfdom is not, and we introduce it![308]

The article's author expressed geopolitical concerns on the project to extend 'British Africa' from the Cape to Cairo, which would have probably led to a new war in North Africa. The cleverness of the 'Britishers' resulted in their politics of alliances, which implied giving 'bribes' to the German emperor allowing Germany to occupy territories of less interest to England, and taking advantage of the temporary weakness of France, 'hampered by her ally, Russia, who can do nothing so long as her robberies on the Pacific and in Manchuria are not consolidated'.[309] A few years later, the new imperial power of Japan would inflict Russia's first defeat in the Pacific, but what is worth stressing here is the geopolitical mindset which characterised *Freedom*'s editors, and paralleled the establishment of 'mainstream' geopolitical scholarship by authors like Halford Mackinder.[310]

Non-European revolutions

In the following years, several notes denounced the enslavement of Chinese workers in South-African mines and the responsibility of all 'Jingoes' for that.[311] Anti-racist concerns were expressed in a correspondence from Johannesburg, which was sent to Kropotkin in 1905 (in a letter which also invited him for a tour of lectures in South Africa, never realised), and commented on by Marsh in a letter he sent to the Anarchist Prince, lamenting the general adhesion to racist and imperialist theories by respectable scholars. 'I have made the extracts from the Johannesburg letter which fills one [with] rage. And the gang of English "scientists" with Brown of Cambridge amongst them – parading under the blood-stained banner of Imperialism!'.[312]

Published in *Freedom*, this South-African note effectively claimed for a major dialogue and understanding between Europeans and natives:

> Besides the interesting white people here, you will find there are the natives, who are much more interesting. Personally, I believe that most of the books about the South African natives have been written in Fleet Street by persons who have never seen a nigger. I have worked in the mine with them, and lived amongst them in Cape Colony, and now I am trading with them; and I can assure you, dear comrade, that I would rather live amongst them than amongst many who call themselves 'civilised'. You can still find amongst them the principle of Communism—primitive Communism. ... I have seen amongst them such brotherly love, such human feelings, such help for one another that are quite unknown between 'civilised' people. And yet my heart is very sore at times when I see them robbed and ill-treated. They must not walk out the pavement, but in the middle of the road. They must not ride in cabs or tram, and in the trains there are separate compartments for them, just like cattle trucks. They must have passes ... and are allowed to live only in the 'location', those Ghettos set aside for them. They are not allowed to be in the streets after 9 p.m., in the land that was once their own—their Fatherland![313]

This early denunciation of what was later called 'Apartheid' is significant because it raised the problem of the difficult relations between white workers overseas and indigenous communities. The fact that several European migrants were sensitised to anarchist propaganda did not automatically imply that they abandoned racist and Euro-centric prejudices. For instance, *Freedom* received several correspondences from socialist and anarchist workers based in Australia from 1886 to 1915, and I never found clear references there to the aboriginal problem, which accordingly indicates a limitation and a contradiction for these early Australian anarchist groups. Yet, all the materials quoted above show that, in the socialist field, most of the early efforts towards anti-racist and anti-colonial thinking came from the anarchists. This also confirms arguments by Lucien Van der Walt about the early South-African anarchist movement, arguing that: 'Local anarchists and syndicalists maintained a principled opposition to racial discrimination and oppression, and a principled commitment to the creation of a multiracial anti-capitalist, anti-statist movement.'[314]

As I already discussed about Ireland, early anarchists likewise paid attention to what is now called 'internal colonialism', and were sensitive to the struggles of nationalities in Eastern Europe and the Caucasus, such as the Finnish, the Polish, the Ukrainians, the Romanians, the Bulgarians, the Serbians, the Georgians, the Armenians and the Greeks, fighting for their autonomy from the empires of Moscow, Vienna and Constantinople. In *Freedom*, these issues were periodically brought to attention, especially through Tcherkesoff's papers.[315] These documents also contributed to the early denunciation by the anarchist geographers of what were later identified as genocides, through their analysis of the first anti-Jewish pogroms in the Russian empire and the first massacres of Armenians in the 1890s.[316] Nevertheless, one of the most significant contributions to anarchist cosmopolitanism of Reclus, Kropotkin and *Freedom*'s editors was their collaboration with activists who came from some of the most distant cultural backgrounds from a European perspective, eventually, the Chinese and Japanese ones.

Early traces of contacts and correspondences can be found in 1890 when an article on 'Anarchism in Japan' questioned the prejudices of non-European peoples who were still widespread in the workers' movement.

> We are apt to consider the Japanese as a semi-civilised race of people to whom Anarchism and Socialism are unknown, and when we proclaim ourselves as Internationalists, many of us never dream of including in the universal brotherhood those islanders of the Far East, precisely because we do not know them, and in an indistinct sort of way perhaps we fear them as a reactionary force. But ... the *Freiheit* of New York recently gave some interesting information on the workers movement in this far-off land.[317]

Freedom and *Freiheit* circulated the call of a Japanese anarchist schoolmistress, Kageame Hidde, writing that: '[She] desires to enter into relation with the revolutionaries of all the countries of the globe and she asks for journals and pamphlets. She will find translators amongst her friends'.[318] The fear of some European unionised workers of competition from cheaper workforces from East Asia had already been criticised by anarchists such as the Recluses in the 1860s and 1870s,[319] and it is not coincidence that contacts between Chinese activists exiled in Europe and members of the Reclus family (especially Paul Reclus) in Paris and the circuits of the New University in Brussels were the origins of the historical Chinese anarchist movement.[320] Likewise, Léon Metchnikoff and the Recluses were influential figures among Japanese radicals and early anarchists at different times.[321] In the case of Japanese anarchist Shūsui Kōtoku (1861–1911), it was accordingly the international renown of Kropotkin, which pushed this activist to get in touch with the Anarchist Prince, writing to him to relate on the progresses of workers' movements in Japan. In February 1907, Kropotkin forwarded to Marsh a letter where Kōtoku argued that 'direct action has begun to [make] tremble the ruling class', complaining about the military repression of a recent miners' strike. Japanese anarchists' tactic was then to propagandise within the Socialist Party to seek the conversion of most of its members. 'A resolution devoted to anti-militarism would be proposed. So, though the Party did not yet give up with parliamentarianism, our libertarian movement is developing rapidly among them.'[322] An abundant literature analyses the figure of Kōtoku as a pioneer of Japanese anarchism, anti-imperialism, and internationalism.[323] These materials show that Kropotkin and *Freedom*'s editors were his first reference persons in Europe, exposing their protagonist role in international and transcultural anarchist and scholarly dialogues.

Kōtoku and his group started a regular correspondence with *Freedom*, describing their commitment to anti-militarism and the harsh repression they often faced, eventually for offenses 'against the Press Law', for which a group of them was arrested. The group included Sanshirō Ishikawa (1876–1956), an activist whom contemporary scholarship considers as strongly influenced by Élisée Reclus.[324] In 1909, a long paper signed 'N' (probably Nettlau) was dedicated to the 'Awakening of the Orient', a phenomenon which firstly challenged what the author deemed an ancient belief which stood beyond the 'attitude of the Fabian Society during the South-African War', that is 'the dogma that civilised peoples are the born leaders and masters of less civilised peoples. How far away these times seem to be in our days when from Morocco to Japan, the Orient is awakening, and the chances of keeping all these peoples under Europe's thumb are simply vanishing into the air'.[325] Interestingly, this wide definition of the Orient anticipated some of Edward Said's arguments about the Orient intended as a cultural construction beyond any clear geographical or historical ground.[326] The paper's main argument was the progressive process of the marginalisation of Europe in the global scenarios which Reclus and Kropotkin already observed in 1890, drawing upon Metchnikoff's remarks on the growing Japanese power and the growing importance of the Pacific scenarios.[327] Moreover, the paper quoted some movements of anti-colonial resistance, including Filipino, Arab and Indian nationalists, which were said to have immediate perspectives for showing that 'enough was enough'. Drawing upon geopolitical arguments analysing the increasing demographical strength of non-European peoples and the geographical spreading of national liberation movements, *Freedom* foresaw elements of the future decolonisation processes. Nettlau's paper argued that anarchists were opposed to efforts 'aiming at the establishment of strong national Parliaments and Governments. But we welcome every blow struck against the prevailing system of colonial conquest'.[328]

However, in Japan's case, *Freedom*'s editors could do little more than denounce the persecutions which anarchists suffered in Japan. Despite this, they tried their best to call international attention on repression in that empire, denouncing the 'Threatened execution of Socialists and Anarchists in Japan' in the first page of the issue of December 1910, which called a public protest meeting, organised to request explanations to the Japanese ambassador in London.[329] Kōtoku was among the arrested, with the allegation of conspiracy against the Emperor's life, and was about to be judged by a tribunal apparently devoid of any democratic guarantee for the defence. In January 1911, the list of the convicted was published in *Freedom*, which insisted on the fact that they were all intellectuals and committed to social betterment, and the need of joining the international support campaign, which was especially intense in the United States.[330] However, this mobilisation was not enough and Kōtoku, together with 11 other comrades, including his partner Sugako Kanno (1881–1911), was hanged on 24 January 1911, at the age of 40. A tribute to him was published in the following issue of *Freedom*, protesting the political nature of the sentence.[331] Shorter correspondences from Japan followed in the journal's successive issues. Recent literature has focused on Kōtoku's work as one of the earliest challenges to Japanese imperialism.[332] Therefore, it is possible to argue that *Freedom* played a global role in constituting the international networks of what Benedict Anderson called anti-colonial imagination.[333]

In *Freedom*'s columns, China was often mentioned regarding the opium affair, which was also used as an argument against the South-African war, as in the case of a 1900 editorial praising Chinese nationalist riots by stating:

> The misdeeds of the missionaries, of the railway engineers, and especially of that gang of plutocrats at Shanghai, who are anxious to begin in China the land grabbing ... which the Chamberlains, the Rhodeses and the titled bankers have carried on in Africa—the misdeeds of all this precious lot have proved to be too much even for Chinese patience. The mass of the Chinese people have had enough of those Catholic and Protestant swindlers who played upon the inexperience of a boy, the Emperor, making him issue edicts against the habits, customs and religion of a civilisation much older than ours; and the old land was aroused by the cry: 'Down with the foreigners!' The Harms-worth-Rhodes's lies-shop at Shanghai is hard at work to induce the British to rush into a war in the Far East. This is precisely what the eastern Rhodeses and the local Harmsworths, Kynochs and Chamberlains want. They want war and annexations in the East.[334]

In 1908, the publication of a Chinese anarcho-communist paper was announced. In its programme, one point was to 'realise internationalism abolishing all the national and racial distinctions', and another one 'to realise absolute equality of men and Azia [sic] women'.[335] In 1913, *Freedom* saluted the first news about Unions' strikes in China, announcing 'the awakening of the Chinese people. Their astonishing revolution, which installed the Republic, is speedily developing economic conditions which will be a lesson to the Western world'.[336] Again, it is worth noting that the concept of 'awakening' was likewise an anarchist's wish for the Western proletariat, therefore, it would be inappropriate to read ethnocentrism in this language. In 1914, a paper on the anarchist movement in China was translated by the Shanghai journal *The Voice of the People*, published in Chinese and Esperanto. In this paper, anti-colonial arguments were put against the 'cruelty' of the Portuguese administration in Macao.[337] In October 1914, a correspondent from Singapore mentioned a forthcoming Chinese anarchist congress,[338] and a 1915 note celebrated Liu Shifu (1884–1915), who had just passed away prematurely, as a pioneer of Chinese anarchism,[339] an assessment which has been confirmed by Arif Dirlik.[340] The idea that the so-called West should learn from Eastern

Freedom

JOURNAL OF ANARCHIST COMMUNISM

VOL. XXV.—No. 262.   FEBRUARY, 1911.   MONTHLY; ONE PENNY.

KOTOKU'S LIFE AND WORK.

[The following sketch of Kotoku and his companions, written just before their execution, has been sent us by one who knew him personally, to whom we are also indebted for the photograph.

DENJIRO KOTOKU was born about 40 years ago in Tosa (Province Shikoku), the son of one who would now be termed a doctor. He came early under the influence of a tutor in his native town who was much advanced for his time, and who already began to question whether "Mitsuhito" rules by divine right, and if the story that his Imperial Japanese Majesty is the "direct Descendant of the Sun" can really be given any credence. As far as I know, Kotoku never attended any middle school, high school, or university, and it is therefore absurd to put "Dr." before his name. A little story will suffice to prove my case. I spoke to him of a mutual doctor friend, and it seems I used "Dr." once too often. Said he, slightly vexed, cigarette in one hand and in the other resting his head: "Why do you omit putting the vocation of others before their name, and never once forget 'Dr.' when you speak of K——?" Of course, it's useless for me to state that I had any intelligent reply to his rather quiet, opportune query.

Kotoku, who strongly resembles a Korean, is gentle, kind, and rather retiring; he is slightly below the stature of the average Japanese, but intellectually so much above them.

Notwithstanding all statements to the contrary, he was editor-in-chief of Japan's most popular paper, published in Tokio, the *Yarakmi-Choha*, which publication can easily bear comparison with the infamous Hearst sheets of this country. Previous to the outbreak of hostilities between Japan and Russia, on account of his anti-war attitude he resigned, and with him Sakai, Nishikawa, and Ishikawa, who with Kotoku form the group in the photograph I enclose. The signatures are in Kotoku's hand-writing.

As the spokesman of the anti-war party, he incurred the displeasure of the Government. From that time on Kotoku and his followers were marked men. Together with the men in the photograph, and others, he started a weekly paper known as the *Heimin-Shimbun* (the Paper of the Common People), which was soon suppressed by the authorities, but was followed by *Hikari* (Light). This publication also was suppressed, and was followed by *Choragen* (Straightforward), which went the way its predecessors did. By this time Sakai, Nishikawa, and Kotoku, in the order mentioned, took their turns and became involuntary boarders of the Japanese Government. "Heimin-Sha" (the People's Publishing House), however, went on turning out pamphlets and books too numerous to mention. The police raided the establishment and confiscated everything, by so doing thinking conservatism is safe and that awful teaching of "revolutionary Socialism and Industrialism" is once for all banished from the shores of the Island Empire.

Kotoku's health was much impaired during his imprisonment (which he, like the rest of them, put to good use by the study of languages), so Dr. Tokidsiro Kato suggested that Kotoku might accompany his eldest son (also a doctor) to America, and later to Europe, at the expense of Dr. Kato. He accepted, went to America, and got acquainted with a new phase of political and economic ideas. During his stay in San Francisco the awful earthquake and fire came between him and the plans of the Katos, so they returned to Japan.

After his return, strong in health, with the support of a rich young fellow countryman (but who later, so it is said, turned traitor and police spy), he started the daily *Heimin-Shimbun*, with a great circulation. For his unquestioned ability as a writer (some of his contemporaries call him the most poetic writer of modern Japan), his sincerity, honesty, and all that is essential to make a good and true man, is acknowledged even by his enemies. But the daily *Heimin-Shimbun* had a short life, and went the same way by the same methods as its predecessors.

From Tokio, the capital of Japan, to the smallest hamlet, Kotoku went about preaching the gospel of human emancipation, fearlessly and straight. No Divinity, said he, would look on and allow a state of things in which the children he created in his own image may starve or be sold into *yoshiwaras* (brothels).—And here let me bring to your notice that after the Russo-Japanese War, Japanese women were actually sold *by weight* for the purpose of prostitution in China. Incredible as it may seem, it is said that during the war, in the city of Tientsin, North China, in the Japanese concession, women of the Red Cross Society, with the regalia and emblem on them, prostituted themselves for the benefit of that organisation.—No person of authority, Kotoku said, has any claim to divinity if he be blind to the poverty and depravity of his immediate surroundings.

Is it any wonder that Kotoku became more and more the thorn in the flesh of the governing class, and at all costs had to be made impossible, no matter how low and despicable the means by which to accomplish their infernal purpose.

I understand that early in August, while already on board a ship bound for America, whence he intended to go to Europe, to attend the International Socialist Congress in Copenhagen, he was arrested, charged, the gods may know with what not, and the result is too well known to enter into details about. I have written to about eight people in various stations of life, asking for details concerning the affair; but it is rather doubtful whether I will get the desired information, in view of the terrible persecutions of people with progressive ideas and the

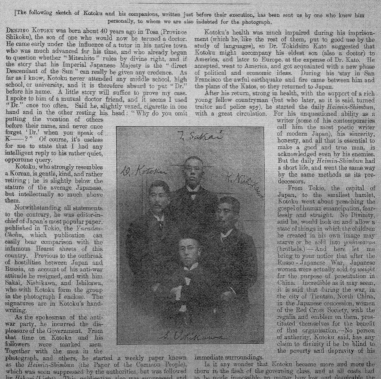

Figure 4.3 The *Freedom* cover dedicated to the Japanese anarchists executed
Freedom, February 1911

experiences is very significant to the fact that being a global hub for anarchism at the Age of Empire meant giving up ethnocentric pretentions on what the world centre of social struggles ought to be.

In the years after 1900, increasing attention was paid to Latin America, with frequent reports on the social struggles carried on in the countries where anarchist and anarcho-syndicalist movements had surfaced earlier, such as Argentina, Brazil, and Cuba. Nevertheless, most of *Freedom*'s attention was called by the 1910 Mexican Revolution. Anarchist historiography considers it as the first great revolution of the twentieth century, arguing that its neglecting was mainly because it occurred outside Europe and was in great part driven by peasants and the indigenous.[341] Within this revolution, the libertarian tendencies led by the Flores Magón brothers (among the inspirers of Emiliano Zapata) are considered among the primary examples of transnational anarchism, given the Magóns' strong international connections and their trans-border activities between Mexico and the United States.[342] An exemplary polemic concerned the French journal *Les Temps Nouveaux*: Jean Grave had warned to be wary of the Flores Magóns' 'Liberal Party', guessing that their model of revolution was substantially extraneous to the anarchist tradition. This triggered polemics with other sectors of the movement, more supportive of Mexican revolutionaries, which are considered today to reveal the 'ethnocentrism' of some anarchists at that time, and which were finally stopped by the respected voice of Kropotkin, who intervened to contradict Grave's opinion, compelling him to apologize publicly with the Flores Magóns.[343] What is important in Kropotkin's arguments are his remarks on the fact that different models of revolution were possible and that, before criticising a movement of peasants and Indians, it was necessary to understand it. Kropotkin complained about the fact that 'unluckily, the 90% (or perhaps the 99%) of the anarchists do not conceive the revolution in other forms than barricade combats, or Garibaldian triumphal expeditions'.[344] For Kropotkin, a peasant's revolution can be made through slower processes, including advances and retreats, than the classical Jacobin myth of a revolution taking the King's palace or the romantic idea 'of those young Italians and Frenchmen who know the revolution from the poems'.[345] What Kropotkin questioned here was not only an urban and Jacobin model of revolution but first and foremost an idea of revolution only based on European insurrectionary imagination. Evoking the importance of the peasants' movements in 1904–1905 in Russia, Kropotkin concluded that what happened overseas could again be of inspiration for workers' movements in Western Europe.

For these reasons, *Freedom* constantly covered the facts in Mexico and published papers by Ricardo Flores Magón (1874–1922). In May 1909, the first appeal from Mexican revolutionaries appeared[346] and, in 1910, *Freedom* published the announcement of the revolt against the oligarchic regime of Porfirio Diaz.[347] In April 1911, a paper from William C. Owen (1854–1929), North-American collaborator of the Flores Magón brothers, celebrated the ongoing events as a 'social revolution'.[348] In the following month, *Freedom* reproduced an appeal signed by Ricardo Flores Magón to North-American workers, already released by De Cleyre's and Goldman's journal *Mother Earth*, in compensation of 'the ominous and disgraceful silence of the English press over the revolution in Mexico'.[349] In 1911, a second appeal clarified that the label of the Magóns organisation should not be confused with European political formations having the same designation (as, apparently, anarchists like Grave had done). 'It must be understood that the Liberal Party, as it is called in Mexico, stands for a genuine Social Revolution.'[350] Therefore, what is clear is that journals like *Mother Earth* in the United States and *Freedom* in Britain endorsed Mexican revolutionaries from the beginning, despite the fact that some European anarchists (accordingly a minority of the movement) remained wary of them as far as they did not display the same symbols and methods of traditional European revolutionary movements. According to Claudio Lomnitz, the label of 'Liberals' was adopted by Mexican anarchists concerning specific aspects of Mexican history which were difficult to understand from Europe's perspective at that time.[351] Something similar occurred again in 1994 after the Neo-Zapatistas' rising in Chiapas, characterised by slogans which seemed not to match any 'classical' line of thinking within Anarchism, Marxism or Social-Democracy.[352]

Significantly, some *Freedom* writings insisted on the involvement of 'natives' in the Mexican social movement,[353] revealing an understanding of the importance of the indigenous element in countries like Mexico, increasingly highlighted by contemporary scholarship on social movements in Latin America.[354] In 1912, a paper by Ricardo Flores Magón, translated by *Regeneración*, provided sketches of the daily life of humble Mexican *peones* and the harassment they suffered from civil and military authorities.[355] Frequent correspondences from that journal were published in *Freedom* in the following years, including an enthusiastic review of the Liberal Party's pamphlet *Land and Liberty*, a motto which recalled the nineteenth-century Russian movement *Zemlia i volia* and was later adopted by peasant movements throughout the world. In 1914, *Freedom* saluted the liberation of the Flores Magóns who had been imprisoned for a period in the United States.[356] In 1915, *Freedom* published the Revolutionary Manifesto by Emiliano Zapata and still referenced the issue of the natives.[357] As recent literature shows, the legacy of the Flores Magóns and the Mexican Revolution is still an important issue for de-colonial movements today in Mexico, Latin America and worldwide.[358] The efforts of early European anarchists, eventually strongly influenced by Kropotkin, to consider different ways of conceiving revolutions, can still provide contributions to ongoing debates on decolonising social movements and social sciences.

A temporary collaborator of *Freedom* in 1907 and 1908, Guy Aldred (1886–1963), is considered as one of the most significant figures of anarchist anti-colonialism in Britain.[359] A young and very ardent revolutionary, Aldred was jailed in 1909 for 'press crime' after he had volunteered to print the journal *The Indian Sociologist* edited by Paris-based Indian nationalist

Shyamji Krishnavarma (1857–1930), whose publication had been forbidden for alleged instigation of political assassination, in a period characterised by attacks on British officers by Indian anti-colonial activists. In his self-defence before the court, Aldred expressed all the complexity of the anarchist debates about anti-colonial nationalist movements. On the one hand, Aldred claimed his support to *The Indian Sociologist*.

> It was ... an organ of freedom, and of political, social, and religious reform. It was edited by Krishnavarma from Paris, and was published for the express purpose of advocating what was called Indian independence, and in furtherance of the Indian Nationalist movement. It was patent, as far as the pages of the paper were relevant to the case, that there was preached ... doctrines intended to bring about the absolute subversion of the Government of His Majesty in the Empire of India, and advocating and urging ... to take all means to throw off what was called the alien yoke.[360]

Conversely, talking as the 'defendant' in the third person, Aldred expressed a series of reservations about nationalism, also when it was an anti-colonial one.

> In his opinion, the workers had nothing to gain as an International oppressed class from identifying themselves with the cause of Indian Nationalism. He remarked, however, that it was the duty of the English military rank and file to refuse to bear arms equally against the Indians, the Egyptians, and the class from which they (the military) were recruited at home.[361]

After his release in 1910, Aldred continued intense activism by setting a committee for the release of convicted Indian nationalist Vinayak Savarkar (1883–1996), he worked in a Ferrer School, edited feminist papers and, finally, promoted anti-militarist insubordination during WWI, spending another two years in jail. For that, according to Nicholas Walter, Aldred 'should be remembered as one of the heroes of the resistance to the First World War'.[362] Meanwhile, Aldred distanced himself from the *Freedom* group considering that they were a 'close corporation',[363] and, quite surprisingly, that 'their anarchy was mere Trade Union activity'.[364] In any case, if he had manifestly argued with activists such as Marsh and Keell, Aldred maintained a correspondence with Nettlau, discussing with him republications of Bakunin materials which were to involve *Freedom*, at least until 1911,[365] and continued also writing to Keell, albeit often in polemical notes, until 1928.[366] Despite these arguments, *Freedom* was actively committed to collect economic contributions for Aldred during his imprisonment between 1909 and 1910.[367] According to Ole Birk Laursen, Aldred's open support of Indian nationalists was somewhat unusual among British anarchists, who were rebutted by the extreme nationalism of many Indian patriots.[368] In any case, the sources mentioned above show that the idea of supporting national liberation when it could be associated with programmes of social liberation, opposing at the same time xenophobic and provincial nationalisms, was generally shared among early anarchists. Scholarship on anarchism and Indian liberation has argued that anarchism has the potential to provide non-nationalist models to de-colonisation movements.[369]

In the first 30 years of *Freedom*, the main focus of anti-imperial critiques was Africa, but the journal also published reports on India, matching Reclus's arguments considering that Indian decolonisation would have happened as an effect of Indians' demographical strength combined with their increasing class consciousness. In a paper deriding missionaries' reports on India, an author signing with the nickname 'Gaelic American' compared the situation in India at that moment with the eve of the American Revolution, arguing that, while 'the "unfitness for self-government" is a cant of the English ... the educated classes in India are disaffected and want to get rid of English rule, and they constitute ... a most formidable force'.[370] The more recently colonised Indo-Chinese peninsula was the object of a paper by Carlyle Potter, who equally matched Reclus's arguments on European colonisation as the disorganisation of local communities, which were often organised in more libertarian and egalitarian ways than the conquerors' society. For Potter, the result was that, in Burma, 'a free, happy and harmonious people are undone, ruined and crushed by our soul-destroying and brutal government as commerce'.[371] The intersection of all these criminal acts ongoing at different times and in different regions of the world was a leitmotif for the anarchist journal. First, the interests of the grand capital were denounced for a plurality of cases summarised in one of Kropotkin's papers:

> In all the wars of the last quarter of a century we can trace the work of the great financial houses. The conquest of Egypt and the Transvaal, the annexation of Tripoli, the occupation of Morocco, the partition of Persia, the massacres in Manchuria, the massacres and international looting in China during the Boxer riots, the wars of Japan—everywhere we find great banks at work.[372]

Second, *Freedom* activists were keen to break commonplace underestimations of extra-European workers' movements which were still widespread among European proletarians, as stated by a South-African correspondent: 'Any socialist of quasi-socialistic movement which totally ignores the larger action of the workers—the coloured and native—will vie in absurdity with the mythical three tailors of Tooley Street'.[373] Denouncing the invention of a Black Peril in British colonies, [374] *Freedom* wished that 'someday the natives may set about remedying their own grievances, then there will be a terrible reckoning'.[375]

It is possible to conclude that *Freedom* was one of the first European journals which radically challenged racism and colonialism, upsetting ironically colonialist commonplaces such as 'the White Man's Burden' (i.e. his alleged duty of civilising the others) by opposing the concept of 'the Black Man's Burden'[376] in European colonies, that is, natives' mandatory submission to the oppressor. If this process was not devoid of problems and contradictions, for instance in the case of the correspondences from Australia (and other settler colonies) mentioned above, it stands clear that anarchists pioneered the progressive awareness of colonial crimes in European workers' milieus. They fully recognised the relevance of social and anti-colonial struggles of different peoples and worked for spreading anti-racist and cosmopolitan views. If the relation between anarchism and nationalist anti-colonialism was complex and not always easy, one does not find, in *Freedom*, arguments such as Marx's, an author whose justifications of colonialism as a necessary step towards 'modernisation' and industrialisation of countries like India, was harshly criticised by Said.[377] Given the prestige, influence and direct intervention of Reclus, Kropotkin and fellow scholars in the journal, it is also possible to conclude that geography was more than relevant for this positioning.

Notes

1   Becker, 'Notes on *Freedom*'; Varengo, *Pagine anarchiche*.
2   Bantman, *The French Anarchists*; Di Paola, *The Knights Errant*.
3   Ferretti and Minder, *Pas de la dynamite*; McKay, 'Kropotkin'; Walter and Becker, 'Introduction'.
4   Ferguson, 'Anarchist women', 710.
5   MacLaughlin, *Kropotkin*; Berneri, *Journey*.
6   Hulse, *Revolutionists*, 10.
7   Kropotkin, *Act for Yourselves*, 116.
8   Bantman, *The French Anarchists*, 33.
9   Wilson, *Anarchist Essays*, 81.
10   IISH, Alfred Marsh Collection, 118, Kropotkin to Marsh [1905].
11   Becker, 'Notes', 8.
12   IISH, Alfred Marsh Collection, 109, Kropotkin to Marsh, 20 June 1905. GARF, 1129, 2, 2161, 4 letters from Rocker to Kropotkin, 1906–1915.
13   Freedom Press, *Our First Centenary*, 1986, 17.
14   Walter and Becker, 'Introduction', 7.
15   Walter and Becker, 'Introduction', 7–8.
16   GARF, 1129, 2, 2495, Wilson to Kropotkin, 2 January 1884.
17   Walter and Becker, 'Introduction', 7.
18   GARF, 1129, 2, 2495, Wilson [to the Freedom group], 16 May 1889.
19   GARF, 1129, 2, 2495, Wilson [to the Freedom group], 16 May 1889.
20   Manfredonia, *Les anarchistes*.
21   GARF, 1129, 2, 2495, Wilson to Kropotkin, 19 July 1905.
22   GARF, 1129, 2, 2495, Wilson to Kropotkin, 19 July 1905.
23   GARF, 1129, 2, 2495, Wilson to Kropotkin, 22 October 1905.
24   GARF, 1129, 2, 2495, Wilson to Kropotkin, 18 June 1906.
25   GARF, 1129, 2, 2495, Wilson to Kropotkin, 5 April 1917.
26   Ferretti, 'Anarchist geographers and feminism'.
27   Kinna, *Kropotkin*, 64.
28   Kinna, *Kropotkin*, 64.

29   GARF, 1129, 2, 1308, Scott Keltie to Kropotkin, 18 February 1878.
30   GARF, 1129, 2, 1308, Scott Keltie to Kropotkin, 14 August 1878.
31   Kinna, *Kropotkin*, 65.
32   'Meeting to welcome Vera Figner'. *Freedom*, July 1909, 53.
33   CWAC, 716/84, Kropotkin to Skilbeck, 3 December 1909.
34   Kinna, *Kropotkin*, 66.
35   Kinna, *Kropotkin*, 68.
36   Varengo, *Pagine anarchiche*, 99.
37   Freedom Press, *Our First Centenary*, 1986, 5.
38   Becker, 'Notes'.
39   IISH, Alfred Marsh Collection, 118, Kropotkin to Marsh [1905].
40   Freedom Press, *Our First Centenary*, 23.
41   Di Paola, 'Marie Louise Berneri'; Sacchetti, *Eretiche*.
42   University College London, Watson Library, Karl Pearson Papers.
43   'Sex and Socialism'. *Freedom*, April 1886, 27.
44   'Sex and Socialism'. *Freedom*, April 1886, 28.
45   'Sex and Socialism'. *Freedom*, April 1886, 28.
46   Brun, 'Élisée Reclus'; Ferretti, 'Anarchist geographers and feminism'.
47   Hinely, 'Charlotte Wilson', 22.
48   Varengo, *Pagine anarchiche*, 100.
49   P. Kropotkin, 'Domestic slavery'. *Freedom*, July 1891, 47.
50   P. Kropotkin, 'Domestic slavery'. *Freedom*, July 1891, 48.
51   P. Kropotkin, 'Domestic slavery'. *Freedom*, July 1891, 48.
52   'The Propaganda Reports', *Freedom*, October 1891, 78.
53   'The Match Girls' Strike'. *Freedom*, August 1888, 89.
54   'The Match Girls' Strike'. *Freedom*, August 1888, 89.
55   'The Women of the Commune'. *Freedom*, April 1888, 75.
56   'Women in Italian rice fields'. *Freedom*, January 1892, 5.
57   'The Women's movement in Russia'. *Freedom*, April 1897, 94.
58   'The marriage controversy'. *Freedom*, October 1888, 2.
59   'Notes'. *Freedom*, May 1897, 38.
60   'Notes'. *Freedom*, April 1889, 20.
61   A. Henry, 'The woman question and state socialism'. *Freedom*, March 1892, 23.
62   A. Henry, 'The woman question and state socialism'. *Freedom*, March 1892, 23.
63   F.S. Merlino, 'The woman question and state socialism'. *Freedom*, March 1892, 24.
64   'Woman and Christianity'. *Freedom*, August 1895, 35.
65   'Notes'. *Freedom*, August 1897, 54.
66   Jordan and Sharp, *Josephine Butler*; Regard, *Féminisme et prostitution*.
67   'Woman's Freedom'. *Freedom*, May 1898, 27.
68   Haaland, *Emma Goldman*; Goldman, *Living my Life*.
69   'Voltairine De Cleyre's tour in Scotland'. *Freedom*, November 1897, 69.
70   T. Claramunt, 'Tenderness'. *Freedom*, November 1897, 74.
71   'The Communist Anarchists and Women'. *Freedom*, December 1901, 70.
72   Red Rose, 'Where women fail'. *Freedom*, July 1907, 42; 'Free Love', *Freedom*, August 1908, 54–55.
73   'Domestic slaves'. *Freedom*, February 1908, 10–11; 'Woman's misery, and the way out', *Freedom*, April 1913, 25; 'Woman's burden" *Freedom*, June 1913, 45.
74   'Woman's work for human freedom'. *Freedom*, August 1908, 57.
75   'Woman in revolt'. *Freedom*, May 1908, 29.
76   L. Parsons, 'The eleventh of November 1887'. *Freedom*, November 1912, 85.
77   Élie Reclus, 'Woman the creator of civilisation'. *Freedom*, November 1910, 86.
78   L. Gair Wilkinson, 'Woman's freedom'. *Freedom*, June 1913, 46; July 1913, 55.
79   Ferguson, 'Anarchist women', 716.

80 IISH, Alfred Marsh Collection, 121, Kropotkin to Marsh, 22 February 1906.
81 IISH, Alfred Marsh Collection, 158, Kropotkin to Marsh, no date.
82 GARF, 1129, 2, 1368, A. Cobden-Sanderson to P. Kropotkin, 3 May [no year].
83 GARF, 1129, 2, 1368, A. Cobden-Sanderson to P. Kropotkin, 26 June [no year].
84 GARF, 1129, 3, 216, letters from A. Cobden-Sanderson to S. Kropotkin, 1912–1926.
85 GARF, 1129, 2, 1368, A. Cobden-Sanderson to P. Kropotkin, 5 June [no year].
86 Nettlau, *Eliseé Reclus, vol. II*, 246.
87 Reclus, *Correspondence, vol. III*, 187.
88 NLS, Patrick Geddes Papers Ms 10564, A. Cobden-Sanderson to P. Reclus, 21 May [1896].
89 NLS, Patrick Geddes Papers Ms 10564, A. Cobden-Sanderson to P. Reclus, 3 January [no year].
90 Cobden-Sanderson, 'Élie and Élisée Reclus', 43–46.
91 National Library of Ireland, Department of Manuscripts (hereafter NLI), MS 49,981/10, Biographical Introduction.
92 NLI, MS 49,981/10, Biographical Introduction.
93 Heath, 'Nannie Florence Dryhurst'.
94 N.F. Dryhurst, 'How I met Kropotkin', *Freedom*, January 1931, 2.
95 NLI, MS 49,981/10, Biographical Introduction.
96 NLI, MS 49,981/10, Journals of George Sturt, 8 May 1902.
97 NLI, MS 49,981/10, Journals of George Sturt, 8 May 1902.
98 NLI, MS 49,981/10, Biographical Introduction.
99 IISH, Nettlau Papers, 3083, Dryhurst to Nettlau, 13 July 1896.
100 IISH, Nettlau Papers, 3083, Dryhurst to Nettlau, 13 July 1896.
101 'The International School'. *Freedom*, March 1891, 19.
102 Bantman, *The French Anarchists*, 69.
103 IISH, Lucien Descaves Papers, L. Michel 203, Dryhurst to Michel, 5 June 1903.
104 Startt, 'The evolution'.
105 Tcherkesoff, *Pages*.
106 V. Tcherkesoff, 'Very scientific, but plagiarism. Considering two Manifestoes'. *Freedom*, May 1900, 22–23.
107 F. Tcherkesoff, 'Death of Mrs. N.F. Dryhurst', *Freedom*, December 1930, 2.
108 NLI, MS 49,981/10, Biographical Introduction.
109 John, *War*, 86.
110 GARF, 1129, 2, 1856, Nevinson to Kropotkin, 23 June 1899, 27 October 1902, plus three letters without year.
111 John, *War*, 90.
112 Nevinson, *Fire*, 52.
113 Nevinson, *Fire*, 53.
114 H. Nevison, [Message from Henry Nevinson], *Freedom*, May 1930, 1.
115 NLI, MS 49,981/10, Biographical Introduction.
116 NLI, MS 49,981/13, The Appeal from the Women of Georgia.
117 NLI, MS 49,981/10, Biographical Introduction.
118 It is worth noting that, at that time, the word 'races' was used as a synonym for 'peoples', as is also shown by studies on scientific literature in Latin languages (Reynaud-Paligot, *La république raciale*; La Vergata, *Colpa di Darwin?*). Identifying racism through the mere use of this term would therefore be an act of anachronism. Interestingly, in the Dryhurst archives, a printed proof of the document quoted in this note presents the word Races corrected as Peoples.
119 NLI, MS 49,981/13, The subject Peoples International Committee.
120 NLI, MS 49,981/13, The subject Peoples International Committee.
121 Grave, *Patriotisme et colonisation*.
122 Dryhurst, 'Nationalities', v.

123   Dryhurst, 'Nationalities', vi.
124   Bey, 'Egypt's demand'.
125   NLI, MS 49,981/10, Biographical Introduction.
126   De Claparède, *L'évolution*.
127   NLI, MS 49,981/10, Biographical Introduction.
128   'International notes', *Freedom*, April 1906, 12.
129   IISH, Nettlau Papers, 373, Dryhurst to Nettlau, 13 November 1907.
130   British Library, MS Add. 46,362.
131   NLI, MS 49,981/16, Copies of letters and notes in relation to Nannie Dryhurst.
132   GARF, 1129, 2, Dryhurst to Kropotkin, 10 August 1906.
133   GARF, 1129, 2, Dryhurst to Kropotkin, 29 January 1909.
134   GARF, 1129, 2, Dryhurst to Kropotkin, 17 May 1909.
135   GARF, 1129, 2, Dryhurst to Kropotkin, 19 November 1909.
136   GARF, 1129, 2, Dryhurst to Kropotkin, 31 March 1908.
137   GARF, 1129, 2, Dryhurst to Kropotkin, 29 October 1908.
138   John, *War*, 164.
139   NLI, MS 49,981/10, Biographical Introduction.
140   McCoole, *Easter Widows*, 200.
141   NLI, MS 49,981/10, Biographical Introduction.
142   McCormack, *Roger Casement.*
143   NLI, MS 13,073/46vii, Roger Casement Papers, Dryhurst to Casement, 15 May 1915.
144   GARF, 1129, 2, Dryhurst to Kropotkin, 9 January 1915.
145   GARF, 1129, 2, Dryhurst to Kropotkin, 26 July 1916.
146   GARF, 1129, 2, Dryhurst to Kropotkin, 2 April 1917.
147   John Morrison-Davidson (1843–1916) was a Scottish socialist historian also involved in *Freedom* meetings.
148   IISH, Alfred Marsh Collection, 158, Kropotkin to Marsh, no date.
149   NLI, Department of Manuscripts, Ms 25,639(59), Kropotkin to Yeats, 5 July 1904.
150   Avakumović and Woodcock, *The Anarchist Prince* 223.
151   Balinisteanu, *Violence*, 16.
152   Balinisteanu, *Violence*, 17.
153   Landuyt and Lernout, *Joyce's Sources.*
154   'Parliamentarianism and revolt' *Freedom*, March 1887, 21–22.
155   'Parliamentarianism and revolt' *Freedom*, March 1887, 22.
156   'Law and Order in Ireland', *Freedom*, October 1886, 8.
157   'Law and Order in Ireland', *Freedom*, May 1887, 32.
158   'Law and Order in Ireland', *Freedom*, September 1887, 48.
159   'Spontaneity'. *Freedom*, August 1887, 43.
160   'Home Rule and after. Impression of an English anarchist in Ireland'. *Freedom*, November 1887, 55.
161   'Home Rule and after. Impression of an English anarchist in Ireland'. *Freedom*, November 1887, 55.
162   'Coercion and revolt in Ireland'. *Freedom*, November 1887, 56.
163   'Face to face with the facts' *Freedom*, December 1887, 57.
164   Ferrari, *Primo Maggio.*
165   'The Irish Martyrs of November. An anniversary and a parallel'. *Freedom*, December 1887, 60.
166   'Ireland for the Workers', *Freedom*, May 1889, 21.
167   'Ireland'. *Freedom*, January 1888, 61.
168   Ferretti, 'Political geographies'; Ferretti, 'Revolutions and their places'.
169   'Ireland'. *Freedom*, June 1888, 84.
170   'Song of Rebellion. A voice from Ireland'. *Freedom*, November 1888, 7.

171 'The propaganda'. *Freedom*, December 1889, 54.
172 'The Belfast Group'. *Freedom*, May 1893, 28.
173 'The Irish split'. *Freedom*, January 1891, 1.
174 'Home Rule and Rome Rule'. *Freedom*, February 1891, 10.
175 'Reports'. *Freedom*, June 1891, 46.
176 Carroll and King, *Ireland*; Kearns, Meredith and Morrissey, 'Spatial justice'.
177 'The Year 1892'. *Freedom*, January–February 1893, 2.
178 'The joys of National School Teachers in Ireland'. *Freedom*, January–February 1893, 5.
179 'Ireland's commemoration. *Freedom*, April 1898, 20.
180 'Ireland, Ireland'. *Freedom*, May 1900, 17.
181 Bantman, *The French anarchists*, 37.
182 'The Sinn Fein movement'. *Freedom*, July 1906, 23.
183 G. Aldred, 'The failure and farce of parliament'. *Freedom*, May 1908, 30.
184 The Irish Rebel, 'The enemies of socialism'. *Freedom*, June 1908, 42.
185 Kinna, *Kropotkin*.
186 'Labour in Irish History by James Connolly'. *Freedom*, May 1914, 35.
187 Nevin, *James Connolly*.
188 'Sidelights on social subjects'. *Freedom*, April 1915, 27.
189 'Repression in Ireland'. *Freedom*, July 1915, 49.
190 'The Tragedy of Ireland'. *Freedom*, May 1916, 37; 'Why Ireland revolts'. *Freedom*, May 1916, 38; 'The Martyrs of Ireland'. *Freedom*, July 1916, 43; 'Conscription for Ireland'. *Freedom*, September 1916, 1.
191 Ferretti, 'Revolutions and their places'.
192 Heath, 'Alfred Marsh'.
193 'Death of Alfred Marsh'. *Freedom*, November 1914, 83.
194 Ferretti, Barrera, Ince and Toro, *Historical Geographies of Anarchism*.
195 IISH, Alfred Marsh Collection, 17, Kropotkin to Marsh, 1895.
196 GARF, 1129, 2, 1700, Marsh to Kropotkin, 6 June 1902.
197 McKay, 'Kropotkin, Woodcock'.
198 IISH, Alfred Marsh Collection, 34, Kropotkin to Marsh, 28 July 1897.
199 IISH, Alfred Marsh Collection, 74, Kropotkin to Marsh, 15 October 1902.
200 GARF, 1129, 2, 1700, Marsh to Kropotkin, 12 December 1912.
201 IISH, Alfred Marsh Collection, 30, Kropotkin to Marsh, 4 November 1896.
202 IISH, Alfred Marsh Collection, 148, Kropotkin to Marsh, 31 December 1909.
203 IISH, Alfred Marsh Collection, 27, Kropotkin to Marsh, 8 August 1896.
204 GARF, 1129, 2, 1700, Marsh to Kropotkin, 12 September 1903.
205 GARF, 1129, 2, 1700, Marsh to Kropotkin, 19 October 1909
206 GARF, 1129, 2, 1700, Marsh to Kropotkin, 22 March 1913.
207 GARF, 1129, 2, 1700, Marsh to Kropotkin, 16 December 1909.
208 GARF, 1129, 2, 1700, Marsh to Kropotkin, 11 February 1911.
209 GARF, 1129, 2, 1700, Marsh to Kropotkin, 23 April 1910.
210 GARF, 1129, 2, 1700, Marsh to Kropotkin, 7 May 1910.
211 Becker, 'Notes'.
212 Levy, 'Foreword'; Turcato, *Making sense.*
213 'Reason-Worship'. *Freedom*, January 1888, 63–64.
214 P. Kropotkin, 'Co-operation. A reply to Herbert Spencer'. *Freedom*, December 1896, 117 (and following numbers).
215 P. Kropotkin, 'Herbert Spencer.' *Freedom*, February 1904, 7.
216 H. Spencer, 'Slavery, the outcome of state socialism'. *Freedom*, June 1903, 33.
217 J. Le Vagre, 'Darwinism and the Revolution'. *Freedom*, March 1891, 17.
218 Avakumović and Woodcock, *The Anarchist Prince*, 211.
219 'Consolations of North Pole expeditions'. *Freedom*, January 1897, 4.
220 Goldman, 'The tragedy'.

221   IISH, Alfred Marsh Collection, 108, Kropotkin to Marsh, 6 May 1905.
222   Volin, *The Unknown Revolution*.
223   IISH, Alfred Marsh Collection, 109, Kropotkin to Marsh, 20 June 1905.
224   GARF, 1129, 2, 1700, Marsh to Kropotkin, 12 July 1911.
225   Kinna, *Kropotkin*.
226   Becker, 'Notes', 21.
227   GARF, 1129, 2, 1700, Marsh to Kropotkin, 7 March 1903.
228   GARF, 1129, 2, 1700, Marsh to Kropotkin, 21 April 1913.
229   GARF, 1129, 2, 1700, Marsh to Kropotkin, 28 June 1913.
230   'Death of Alfred Marsh'. *Freedom*, November 1914, 83.
231   Ferretti, 'Teaching'; Pelletier, *Géographie et anarchie*; Springer, *The Anarchist Roots*.
232   Hirsch and Wan der Walt, *Anarchism*.
233   Anderson, *Under Three Flags*; Anderson, *Life Beyond Boundaries*.
234   Reclus, *Correspondance, vol. II*, 337.
235   Reclus, *Correspondance, vol. II*, 338.
236   Reclus, *Correspondance, vol. II*, 338.
237   Michel, *Exile*.
238   Reclus, *Correspondance, vol. II*, 338.
239   Hechter, *Internal Colonialism*.
240   Reclus, *HT, vol. VI*, 2.
241   Newsinger, 'Why Rhodes'.
242   Reclus, *HT, vol. VI*, 2.
243   Altena, 'English and Dutch anarchists'.
244   Reclus, *HT, vol. VI*, 3.
245   Reclus, *HT, vol. VI*, 6.
246   Reclus, *HT, vol. VI*, 28.
247   Reclus, *HT, vol. VI*, 28.
248   Reclus, *HT, vol. VI*, 50.
249   Scott, *The Art*.
250   Reclus, *HT, vol. VI*, 70.
251   'Another little war'. *Freedom*, March 1887, 22.
252   Ferretti, 'Arcangelo Ghisleri'.
253   'Another little war'. *Freedom*, March 1887, 22.
254   Alavoine-Muller, 'Introduction'.
255   'Notes'. *Freedom*, June 1888, 82.
256   Harris, *Rebellion on the Amazon*.
257   'Notes'. *Freedom*, June 1888, 82.
258   Romani, *Oreste Ristori*.
259   Hecht, *The Scramble*.
260   'The Commune of Paris'. *Freedom*, March 1889, 13.
261   'The revolt of the English workers in the nineteenth century'. *Freedom*, May 1889, 24.
262   'The workers of South Africa'. *Freedom*, July 1889, 30.
263   L. Holmes, 'The Red Indians and the American Government'. *Freedom*, March 1891, 18.
264   L. Holmes, 'The Red Indians and the American Government'. *Freedom*, March 1891, 18.
265   Ferretti, 'Evolution and revolution'.
266   'Roman Justice'. *Freedom*, December 1891, 91.
267   'Opium culture in India'. *Freedom*, April 1893, 22.
268   'Opium culture in India'. *Freedom*, April 1893, 22.
269   Reclus, *HT, vol. V*, 166.
270   'Manufacturing war'. *Freedom*, September 1893, 60.

271   Gregory, *The colonial present*.
272   Reclus, 'Hégémonie'.
273   Kropotkin, *Anarchism*.
274   'War'. *Freedom*, February 1896, 65.
275   'English atrocities'. *Freedom*, February 1896, 66.
276   'Chivalry and barbarism'. *Freedom*, February 1896, 66.
277   'Telegrams and the truth. *Freedom*, April 1896, 74.
278   'Pity the poor landlords!'. *Freedom*, May 1896, 77.
279   Reclus, 'Léopold de Saussurre'.
280   F. Nansen, 'Christianity and barbarism'. *Freedom*, May 1896, 78.
281   F. Nansen, 'An uncivilised view of civilisation'. *Freedom*, July 1896, 86.
282   'Élie Reclus'. *Freedom*, March 1904, 11.
283   'Eskimos and Communism'. *Freedom*, March 1910, 17.
284   'Savages?'. *Freedom*, November 1897, 70.
285   J. Perry, 'Brutes in unity'. *Freedom*, December 1897, 76.
286   Muthu, *Enlightenment Against Empire*.
287   'International notes'. *Freedom*, April 1898, 23.
288   'Saving the Soudan'. *Freedom*, May 1898, 28.
289   E. Reclus, 'War'. *Freedom*, May 1898, 26.
290   'War'. *Freedom*, November 1899, 77.
291   'War'. *Freedom*, November 1899, 77.
292   'The British workers and the war'. *Freedom*, March–April 1900, 9.
293   'The British workers and the war'. *Freedom*, March–April 1900, 9.
294   'The British workers and the war'. *Freedom*, March–April 1900, 10.
295   'Fabianism Furioso'. *Freedom*, May 1900, 20.
296   G.B. Shaw, 'The man of destiny, written in 1895' *Freedom*, May 1900, 22.
297   IISH, Alfred Marsh Collection, 71, Kropotkin to Marsh, October 1901.
298   IISH, Alfred Marsh Collection, 71, Kropotkin to Marsh, October 1901.
299   IISH, Alfred Marsh Collection, 87, Kropotkin to Marsh, 23 August 1903.
300   British Library, Add Ms 50514, 113–116, Kropotkin to Shaw, 23 August 1903.
301   'International notes'. *Freedom*, March–April 1900, 15.
302   [F. Tarrida] Del Marmol, 'Cuba'. *Freedom*, July 1899, 55.
303   A. Barton, 'To the Filipinos'. *Freedom*, May 1899, 35.
304   'International notes'. *Freedom*, December 1899, 82.
305   M. Twain, 'Imperialism pilloried.' *Freedom*, May–June 1901, 21–22.
306   'The workers of France to the workers of Great Britain'. *Freedom*, July 1901, 29.
307   'To the people'. *Freedom*, December 1901, 69.
308   'One war is over – When the next?'. *Freedom*, June 1902, 21.
309   'One war is over – When the next?'. *Freedom*, June 1902, 21.
310   Sidaway, Mamadouh and Power, 'Reappraising geopolitical traditions'.
311   'The fruits of Imperialism'. *Freedom*, March 1904, 10; 'Jingo *par excellence*'. *Freedom*, October 1904, 38.
312   GARF, 1129, 2, 1700, Marsh to Kropotkin, 7 November 1905.
313   'International notes – South Africa'. *Freedom*, November–December 1905, 38.
314   Van der Walt, 'Revolutionary syndicalism', 33.
315   'Socialist literature for Poland'. *Freedom*, December 1891, 90; W. Tcherkesoff, 'Armenia and Europe', *Freedom*, December 1896, 119; W. Tcherkesoff, 'Greece and Europe'. *Freedom*, June–July 1897, 43–44.
316   Ferretti, Malburet and Pelletier, *Élisée Reclus*.
317   'Anarchism in Japan'. *Freedom*, August 1890, 34.
318   'Anarchism in Japan. *Freedom*, August 1890, 34.
319   Ferretti, 'De l'empathie'.
320   Dirlik, *Anarchism*.
321   Konishi, *Anarchist Modernity*; Willems, 'Contesting imperial geography'.

322   IISH, Alfred Marsh Collection, 132, Kōtoku to Kropotkin, 15 February 1907, enclosed in Kropotkin to Marsh, 23 March 1907.
323   Pelletier, *Kôtoku*; Nothelfer, *Kōtoku*; Tierney, *Monster.*
324   Nozawa, 'Development'.
325   'The awakening of the orient'. *Freedom*, January 1909, 4.
326   Said, *Orientalism.*
327   Ferretti, 'The correspondence'.
328   'The awakening of the orient'. *Freedom*, January 1909, 4.
329   'Threatened execution of Socialists and Anarchists in Japan'. *Freedom*, December 1910, 90.
330   'Japanese Socialists'. *Freedom*, January 1911, 5.
331   'Kotoku's life and work'. *Freedom*, February 1911, 9–10.
332   Tierney, *Monster.*
333   Anderson, *Under Three Flags.*
334   'An urgent need: a labour convention'. *Freedom*, September–October 1900, 37.
335   'A Chinese anarchist paper'. *Freedom*, July 1908, 52.
336   'A Chinese strike'. *Freedom*, January 1913, 1.
337   'Anarchists in China'. *Freedom*, July 1914, 50.
338   'Anarchism in China'. *Freedom*, October 1914, 78.
339   'A Chinese anarchist pioneer'. *Freedom*, October 1915, 78.
340   Dirlik, 'Anarchism'.
341   Shaffer, 'Tropical libertarians'.
342   Lomnitz, *The Return.*
343   Kropotkin, 'Rectification'. *Les Temps Nouveaux*, 27 April 1912, 1–2.
344   Kropotkin, 'Rectification'. *Les Temps Nouveaux*, 27 April 1912, 2.
345   Kropotkin, 'Rectification'. *Les Temps Nouveaux*, 27 April 1912, 2.
346   'To the Revolutionists of all countries'. *Freedom*, May 1909, 37.
347   'The Mexicans rebel'. *Freedom*, December 1910, 91.
348   W. C. Owen, 'Social Revolution in Mexico'. *Freedom*, April 1911, 28.
349   'Appeal of Mexico to American Labour'. *Freedom*, May 1910, 33.
350   'The revolution in Mexico'. *Freedom*, July 1911, 52.
351   Lomnitz, *The Return.*
352   Zibechi, *Dispersing Power.*
353   W. C. Owen, 'Mexico and the United States'. *Freedom*, February 1912, 13; 'The Mexican Revolution, *Freedom*, December 1912, 93–94.
354   Stahler-Sholk, Vanden and Becker *Rethinking.*
355   R. Flores Magón, 'What good is authority?' *Freedom*, May 1912, 38.
356   'The Mexican Rebellion.' *Freedom*, April 1914, 28.
357   'Mexican's Revolutionary Manifesto'. *Freedom*, January 1915, 2.
358   Ellison, 'Banging'; Souza, 'Lessons from praxis'.
359   Laursen, 'Anarchist anti-imperialism'.
360   Aldred, *Dogmas*, 23.
361   Aldred, *Dogmas*, 29.
362   Walter, 'Guy A. Aldred', 85.
363   Aldred, *Dogmas Discarded, Part Two*, 53.
364   Aldred, *Dogmas Discarded, Part Two*, 56.
365   IISH, Max Nettlau Papers, 140, Aldred to Nettlau, 3 October 1911.
366   IISH, Freedom Archives, 435, Aldred to Keell, 1907 to 1928.
367   'Aldred Fund', *Freedom*, December 1909, 96.
368   Laursen, 'Anarchist anti-imperialism'.
369   Ramnath, *Decolonising Anarchism.*
370   'Six million educated Indians'. *Freedom*, May 1909, 34.
371   'Free society in Burma and English rule'. *Freedom*, April 1909, 26.
372   P. Kropotkin, 'Modern Wars and Capitalism'.

373 'South Africa'. *Freedom*, August 1913, 67.
374 G. Harrison, 'The Black Peril in South Africa'. *Freedom*, November 1911, 83.
375 'Creating a "Black Peril"'. *Freedom*, March 1913, 18.
376 'The Black Man burden'. *Freedom*, December 1912.
377 Said, *Orientalism*.

5 Ripples and waves of anarchist writing

Towards humane sciences

Joseph Ishill, the North-American editor of the 1927 recollections on the Reclus brothers mentioned above, had already published similar work in 1923, where he collected tributes to Kropotkin, defining the Anarchist Prince as not only a 'thinker' and a 'rebel' but also a 'humanitarian'.[1] Likewise, Ishill defined the Reclus brothers, together with Godwin, Bakunin, Kropotkin, and Tolstoy as having 'courageously challenged society as being cruel and inhumane [developing] humanitarian ideals for the betterment of mankind'.[2] It is not a coincidence that both Ishill's publications contained recollections form British scholars akin to what has been called 'ethical socialism',[3] such as Havelock Ellis and Henry Salt (for both Reclus and Kropotkin), Edward Carpenter (for Kropotkin), Patrick Geddes (for Reclus), and others. This chapter delves deeper into the analysis of correspondences and other primary sources to investigate Reclus's and Kropotkin's relations with these British figures of scholars defining themselves as 'humanists' or 'humanitarians' such as Richard Heath, Edward Carpenter, Henry Salt and Havelock Ellis, and in an even broader sense, William Morris, Walter Crane, James Mavor, and Patrick Geddes, for the manifold challenges they launched to the intellectual establishment in the construction of more 'humane sciences'. A drive towards 'humane' matters was strictly associated with some specific characteristics of British socialism, including the high importance that education, art and literature had in these milieus.

Important British acquaintances of Kropotkin and the Recluses such as Carpenter, Ellis, Salt (and more indirectly Geddes) participated in a group called the Fellowship of the New Life.[4] Founded in 1882 by the Scottish philosopher Thomas Davidson, the Fellowship had first split in 1884, when a part of its members resigned and founded the Fabian Society, though a certain fluidity and porosity characterised the membership of these associations. Traditional socialist historiography has generally discarded the Fellowship in a very haughty and despising way, considering it as a mystic and utopian 'deviation' from the main road of class struggle. More recent literature has focused on the social relevance of some of these activists' contents, stressing the importance of changing social relations in everyday life by addressing questions related to sex, feelings, companionship, feminism, civil rights, and animal rights as political problems. In particular, Kevin Manton has challenged the traditional 'mould of derision or disregard'[5] which has surrounded the Fellowship, calling for a reconsideration of 'English ethical socialism'. In a book published at the occasion of the exposition 'Anarchy and beauty, William Morris and his legacy' held in London in 2014, Fiona MacCarthy put more emphasis than she had in former works on the connections between some anarchists, some members of the Socialist League, and some members of the Fellowship of the New Life in the common task of 'dismantling of the stultifying structures of society and their replacement by a freer, more equitable and fluid way of life'.[6] In particular, MacCarthy argues that 'the great meeting point between Carpenter and Morris was their joint yearning for simplicity of living'[7] and the pedagogical importance of handcraft, and that 'in his visionary concepts, Kropotkin too was a *News from Nowhere* man'.[8] In a nutshell, these very different scholars and activists shared common concerns in conceiving a revolution which could go beyond mere economic transformations in addressing a wide array of material and psychological aspects of daily life.

The Fellowship of the New Life was also directly participated in by *Freedom* collaborators such as John Coleman Kenworthy (1861–1948), a Christian anarchist inspired by Tolstoy,[9] and Irish anarchist Agnes Henry, who lived in a communitarian house held by the Fellowship[10] prior to lending a part of her house to the *Freedom* printing shop as detailed in Chapter 4. Both Kenworthy and Henry were Morris's correspondents; Kenworthy also invited Morris to lecture in Liverpool in 1890.[11] My arguments are reinforced by sources related to an association founded by Salt in 1891 and actively supported by Carpenter, by the atheist and freethinker Hypatia Bradlaugh Bonner (1858–1935), and by Reclus and Kropotkin, the *Humanitarian League*. This group, which was at the origin of publications such as *Humane Science Lectures* and the journals *Humanity* and *Humane Review*, is still little studied. Nonetheless, in works on the history of civil rights in Britain, the *Humanitarian League* is mentioned as a protagonist of early efforts to fight against 'corporal and capital punishment'[12] and to abolish the death penalty in England. Works by Dan Weinbren also demonstrated the wide outreach of the topics raised by the association beyond its small dimensions, and its opposition to social Darwinism, which remained in an evolutionist theoretical framework [13] like the works of the anarchist geographers. A more recent book by Rod Preece has shown that ideas on animal sensibility and animal rights were not extraneous to wider debates in Victorian England, and eventually in Shaw's works.[14] If humanitarianism is not a straightforward definition for the anarchist movement, it is worth noting that the broad concepts of humanity and humanism as addressed by Salt and his fellows corresponded to some of Reclus's and Kropotkin's concerns. In Emma Goldman's recollections, the term 'humanitarian' was used to define Kropotkin, albeit Nettlau considered Reclus's thinking as reposing 'on a more humanitarian basis'[15] than the Anarchist Prince. Errico Malatesta, who was opposed to Kropotkin in the debates on WW I, defined him as 'the most greatly humane man', considering 'his entire life [as] a labour of love'.[16] This shows how some of the most important international figures of anarchism publicly self-identified with definitions such as 'humane', 'humanitarian' and feelings like love.

These umbrellas of 'humane' concerns could keep together a heterogeneous array of subjects, including civil and gay rights pioneered by authors like Edward Carpenter and supported by *Freedom*, animal rights in the case of Henry Salt's *Humane Review*, struggles against torture and the death penalty and the strife for including humane and ethical concerns in scientific and scholarly work. I include in this chapter unorthodox activists and scholars such as William Morris and Patrick Geddes, for their common concern with popular education and humanitarian endeavours. For that, I advance a two-fold argument, extending the literature mentioned above. First, all these social, humanitarian, and civil endeavours were considered by early anarchists as useful tools to have a say in British public debates, occurring mostly at the level of social mobilisation, periodical publishing and the 'public opinion' rather than at the parliamentary levels. Therefore, they were compatible with anarchist ideas on 'action from below', and all participated in intersectional ideas on social and ethical progress. Second, the drive towards 'humane science' was consistent with anarchist ideas of ethics in scholarly work, challenging mechanical rationalisms and the pretended 'neutrality' of science. As recent scholarship showed, feelings, emotions and sensorial experiences were part of the 'anarchist roots of geography' and ethical stances like 'brotherhood' and 'love for humankind' were likewise the main theme for anarchist propaganda.[17] Finally, it is possible to argue that these authors considered humanitarianism as synonymous with humanism, and that this definition had nothing to do with anthropocentrism, charity or other commonplace understandings of this terminology. As I explain below, these ideas in 'humane science' did not only strengthen the humanistic side of anarchism as expressed by figures like Malatesta, but also anticipated key features of the great intellectual movement known as 'humanistic geographies', which challenged quantitative approaches and technocracy, hegemonic in the discipline in the 1970s.[18]

The most 'humane' collaborator: Richard Heath

Reclus, and Heath's French connection

As anticipated, Richard Heath, a supporter of the *Humanitarian League* and author for the *Humane Review*, was one of the most important British friends and collaborators of both Reclus and Kropotkin. He contributed to introducing them in the 'humane science' circuits, and it is possible to argue that common concerns with some general principles of humanitarian socialism were his main link with the two anarchist geographers, beyond some professional collaborations. Indeed, Heath's philosophical ideas, inspired by a strong belief in the Christian religion, were a far cry from Reclus's and

Kropotkin's anarchism and atheism. Nevertheless, Heath produced an immense amount of correspondence with both Kropotkin and Reclus; unluckily, surviving materials are in both cases incomplete and asymmetrical, as we only have 33 letters sent to him by Reclus (in French), without Heath's answers. In Kropotkin's case, there is the opposite problem: only Heath's letters (more than 100) survive in the Moscow archives, while no public archive holds Kropotkin's letters received by Heath and, until now, it was impossible to locate them in any private collection. However, the amount of the materials, their contexts and the depth of the topics addressed allow following these exchanges and tracing some parallelisms between the Heath–Reclus and the Heath–Kropotkin relationships.

Louise Dumesnil explained the reasons for the life-long friendship between her brother Élisée Reclus and Heath, arguing that the English historian was not only 'one of the best of Élisée's friends in England, but also a raffinate scholar of Edgar Quinet and a humanitarian socialist'.[19] Therefore, they shared common scholarly and cultural interests, as well as general political and ethical concerns. 'Despite the divergence of their opinions, the same feeling of universal solidarity characterised a sincere Christian like Heath and an anarchist-communist like Élisée. They had the same ideal: humans' happiness.'[20] As anticipated above and as confirmed below by Reclus's and Heath's exchanges on the Anabaptist tradition, their common reference to traditions of religious Dissent, for Heath by choice, and for Reclus by his family's tradition, also played a role in building and maintaining this friendship.

After their meeting in 1852 in London, Heath and Reclus remained almost 20 years without having any further contact. In November 1871, Heath read in the newspapers of Reclus's post-Commune sentence and decided to write to him, to express solidarity and support. In his French prison, Reclus only received that letter several weeks later, and responded on 8 January 1872 declaring his happiness in receiving 'your good letter of 18 November ... after a separation of 20 years'.[21] What looks striking in this letter is that it shows a feeling and a general agreement on principles in two men who did not talk for many years and yet they became confidents immediately. Heath had apparently praised Reclus's behaviour in refusing political mediations and abjurations and facing the process with the full dignity and responsibility of his actions. Reclus acknowledged his friend arguing that: 'My great consolation is that I acted following my conscience ... It is sweet to know that you understand me at the point that you are sure that I did not fight for violence, interest or greed of power.'[22] At that time, Reclus' perspective was still to be deported, a situation which he explained to Heath, writing that his destiny was 'similar to [that of] English deportees, with the difference that it is not Australia, but New Caledonia'.[23]

In the immediately following years, the exchanges between the two men were rare or not documented. They became more frequent in the early 1880s concerning Heath's book on Edgar Quinet (1803–1875). Quinet was an important figure for the French left in the nineteenth century. Protestant in a country where Protestantism was often equated to unorthodoxy, republican and 1848-revolutionary, Quinet was a historian at the prestigious *Collège de France*. He had been one of the inspirers of Reclus's generation of French radicals. In January 1880, Heath was seeking the address of Quinet's widow for a forthcoming trip to France, and Reclus put him in touch with the best person for introducing him to Quinet's family and to provide guidance in the Parisian intellectual milieus: 'May I suggest you an excellent introducer, my brother-in-law Alfred Dumesnil, who substituted Quinet at the *Collège de France* and who always remained a close friend of Quinet's family.'[24] Alfred Dumesnil (1821–1894) settled in the Vascoeuil castle in Normandy, former home of famous French historian Jules Michelet (1798–1874), when he became Michelet's son-in-law.[25] Widowed in the 1860s, Dumesnil married Élisée's and Élie's sister Louise Reclus and became a strict collaborator of the Recluses. His position as the editor of Adolphe Lamartine's works and his numerous acquaintances in political and editorial milieus rendered him a precious contact, of which Heath took advantage, starting another long professional correspondence with Dumesnil, and then with Louise after her husband's death. On the occasion of Heath's trip to France, Reclus was also instrumental in giving him 'an introduction letter for Hachette',[26] the historical publisher of the anarchist geographer. Strangely, in Heath's biographical book on Quinet, only Dumesnil was publicly acknowledged, and not Reclus. Some discretion seemed to have inspired Heath in managing these prestigious but sometimes 'embarrassing' acquaintances: before a 1905 obituary that he dedicated to Reclus, Heath never claimed his friendship with him, or with Kropotkin, in his publications, at least to the best of my knowledge.

Heath's relationship with Reclus was characterised by exchanges on deep philosophical questions, and it is possible to imagine that he wrote lengthy letters to Reclus, as he did with Kropotkin. The French anarchist geographer, very busy with his work for the NGU, tried to delay these discussions to more opportune occasions, for instance writing to Heath that: 'If you come to visit me in Switzerland ... we could discuss more widely and seriously of the elevated matters you raise in your letter'.[27] In February 1882, Reclus took the time for responding to some of Heath's remarks which apparently criticised Parisian life.

> It is not because I dislike Paris that I seldom visit Paris ... if you had seen Paris as we saw it when it was sieged, and when the matters on "mine and yours" ... were secondary, if you had enjoyed this fraternity which bonded all us mutually, then you would have lived some moments of that indefinable joy where one is happy of being human and living with humans.[28]

Mentioning the progressive aspects of Enlightenment and the French revolution, Reclus expressed his ideas of cosmopolitanism, which perhaps included implicit blame for some British chauvinism he might have remarked in Heath's last letter.

The place where the human rights have been proclaimed, Paris, is as much worthy as Oxford, where one has just proclaimed the rights of nobles and rich to some more or less authentic instruction believe me. My dear friend, I don't talk as a chauvinist. Very sincerely, I am as much English or Chinese as French in my desire of justice and love.[29]

In a letter of March 1882, Reclus expressed some surprise in understanding that Heath had become a believer, something which clashed with Heath's biography stating that he 'joined the Baptist Church'[30] in 1849; therefore, he was already a believer when he frequented Reclus in 1952. In any case, at that time, it was more likely that the two youths discussed revolutions rather than theology: between 1847 and 1952, Heath not only read the Bible but also works of continental republicans such as Mazzini and Lamartine, at the point that 'he rather revelled in being called a Red Republican',[31] the same political definition that Reclus used for himself then. Indeed, what still kept Reclus and Heath together was their shared non-conformism. Reclus observed that: 'albeit becoming a believer, you remained a free thinker'.[32] Accordingly, in response to new Heath's arguments for religion, Reclus fondly criticised Christianity, saying that the idea of socialism, expressed by the experience of the Commune, wished salvation for everybody and not only for the 'elect'. In these exchanges, it is possible to appreciate the common Protestant origin of Reclus and Heath. Although refusing to become a minister when he was offered that option,[33] Heath remained close to the Baptist Church as an independent thinker. In the context of a theological dispute in which I will not delve deeper, Calvinists and Baptists share some concepts on 'predestination' and on the 'elect'. Comparing socialism with independent Churches (and with a reference to the Methodists accordingly in response to some of Heath's remarks), Reclus concluded that: 'I doubt that these English Methodists, hindered by their mistake [that of being believers], can have a similar free and powerful action than those who got rid of their old prejudices [that is, the atheists]'.[34] For Reclus, socialism gave more concrete guidance than faith.

From Switzerland, Reclus continued to give hints for Heath's sojourns in Paris. His letters reveal that his British correspondent was interested in books on the 1871 Commune and that Reclus praised Heath's works on the English peasants, which would be published in a volume in the following years,[35] for his attention to the daily life and concrete problems of humble people. 'You told me about the humble virtues of the English workers with whom you lived. They are not corrupted by power, intrigues and lies of the government, and of diplomacy.'[36] However, the discussion often shifted to matters of religion, where one has the impression that Heath tried obstinately to convince Reclus of his position, as he did later with Kropotkin. In July 1882, Reclus discussed the links between Christianity and Republics in Mazzini's thinking in the period around 1848: 'I have a rather clear impression of what happened at that time, when I entered in public life for the first time, and I remember well that this reconciliation between republic and Christianity was a result of the confusion that one did at that time between religion and ethics.'[37] Reclus also pleaded for his own belief: 'The anarchist is that who does not recognize any master and does not want to be master for anybody ... the contrary of the one who kneels down before another man who speaks on behalf of God.'[38] Again, another ideal was opposed to Christianity.

Significantly, several of Reclus's letters discuss the topic of women's rights with Heath. These conversations accordingly started after some of Heath's remark on the alleged 'backwardness' of feminine emancipation in France. Reclus replied that: 'You are unjust to French women ... In our workers circles, the question of equality is practically resolved, without contestations.'[39] Reclus's remarks were accordingly too optimistic, but this letter contained an important reply on the issue of Proudhon's misogyny. Pierre-Joseph Proudhon, one of the 'Founding Fathers' of anarchism,[40] was often criticised for his views opposed to women's emancipation. Nonetheless, Reclus claimed that members of the following generation of anarchists (including himself and Kropotkin) harshly condemned Proudhon's views, and always refused the label of 'Proudhonites', due exactly to that controversy. As he wrote to Heath: 'As for the example of Proudhon you present, all socialists fought it, and Proudhon's pages on women are still for us all what most tarnishes the memory of this socialist writer'.[41] In August 1882, the two men had a first exchange about Josephine Butler, whom they both admired for her work in favour of women's dignity, where Reclus declared that feminism was definitively part of anarchism.

> The name of Josephine Butler, which you evoke, inspires my deep sympathy. As I did not share your admiration for other names, like Garfield or Gladstone, that you already quoted in other occasions, I want to stress how much I admire and like this dedicated person who did not fear to expose herself to outrage, to hateful contacts, for rescuing women in difficult situations and defending their dignity against the injustice of laws. I would be happy to help her directly, but indeed the cause I strive for (though worthlessly) comprises among its claims those to which Lady Butler is committed.[42]

Reclus was acquainted with Butler, corresponded with her, and also met her in Switzerland in 1887. In Butler's archives surviving at the London School of Economics, one of Reclus's letter to Butler expressed his great deference towards that figure: '[I wish to] present you the expression of my deep sympathy. Your name is one that I pronounce with emotion, knowing how valorously you have struggled for the cause of human dignity.'[43] This did not prevent Reclus from unchaining his irony on the upper classes he was entitled to meet in his capacity of world-renowned scientist during his meeting with Butler in Switzerland. As he wrote to Heath: 'Unluckily, I could not spend much time with Lady Butler. I arrived during a big dinner, with men in black suits and ladies with colliers and pendants. However, some sincere sentences, some deep gazes were sufficient for me to feel how much Lady Butler is good and honest. I much venerate her.'[44] Nevertheless, Butler's works were still considered as prickly by conformists, as their possible translation to French was giving headaches at the highest political levels in 1889. 'French government, in its great solicitude for the happiness of its citizens, studies the two Butler's works for eight days, in order to stop the principles of a subversive moral.'[45] Therefore, also a common sensibility to early feminism was shared by the two correspondents, one which was not straightforward for a Victorian mystic like Heath.

An important issue under discussion was violence. As Heath's letters to Kropotkin mentioned below confirm, violence was one of the aspects of anarchism and socialism which most frightened Heath, always keen to request explanations from his famous anarchist pen pals. First, Reclus clarified that his critique to 'enemy' ideologies such as Christianity was not informed by rage or resentment: 'You defend Christianity as you understand it; I fight it, *sine ira*, because I understand it in a different way and it seems to me bad as those sewer waters which remained to long time under a field and corrupt it instead of fecundating.'[46] This did not exclude harsh statements such as: 'Between protestant Jesuitism and the catholic one, I don't know what is the worst.'[47] However, finally, anarchism did not mean adopting violence as a programme: 'For revolution, I would never preach violence, and I am sad when friends drag by passion are seduced by the idea of vengeance. But the armed defence of a right is not violence.'[48] During these discussions, Reclus introduced Heath to a Russian author who mostly interested him as correspondence with Kropotkin shows—Leo Tolstoy (1828–1910), one of the rare exponents of Christian anarchism. Though admiring him, Reclus first diverged with Tolstoy on theological matters.

> There is no Book from which Truth spreads: we can come to know it only for external work, for the continual beating of the blood in the arteries, of thinking in the brain. Do you know the recent works of Count Tolstoy? ... I have a deep sympathy for him, but I think that he is wrong, like you, in dividing "The Man's Son" from other men to divinise him, giving to this story a superior value than other human words.[49]

Second, Reclus criticised Tolstoy's idea of passive resistance, arguing that giving up self-defence from the evil was worse than the evil itself, and 'anti-humane'.

Reclus also provided examples very familiar to Heath, a historian of religious movements. 'Despite the horrors of the social war, I side with the Anabaptists, the Jacques, the losers and oppressed of all names, from all nations and times.'[50] The Anabaptists, on which Heath much worked, were also mentioned by Kropotkin among the broad forerunners of anarchism in his entry for the *Encyclopaedia Britannica* on anarchism mentioned on p. 45. The last example provided by Reclus in his later letters was Blaricum in the

Netherlands, a free commune informed by anarchist principles, whose members decided to stop their social experiments after having been violently assaulted by their peasant neighbours. For Reclus, these events were especially detestable because religious fanaticism and social conformism only dictated the hate against that commune, and the commune's dwellers had never harmed anybody in the surroundings. Therefore, Reclus criticised the passive choice of the Blaricum comrades: 'The misadventure of the Blaricum members was sad, but I am one of those who deplore that the comrades did not defend themselves. One curiously facilitates the evil while one let it act without protesting, and on my mind, abandoning the weak to the violence of the strong is betraying the cause. We must resist the evil without hating the evils.'[51] According to anarchists such as Reclus, Kropotkin and Malatesta, the principle of violence had to be rejected, but resistance was a moral duty: for them, resistance should never be confused with hate or indiscriminate violence.

Another frequent matter of the conversations between Heath and Reclus anticipated one of the key topics of *Humanity* and the *Humane Review*, which was the question of animal rights. In his letters, Reclus anticipated some elements of what is today called anti-speciesism, a feature of his thinking which has been highlighted by recent literature.[52] Noticing Heath's sadness for 'the cruelties inflicted to animals', the anarchist geographer did not resist the temptation of an anti-clerical analysis of their causes: 'This Christian idea which refuses them a soul and considers them pure machines played a big role in the abominations which are accomplished everyday against poor animals.'[53] Human and non-human social problems were compared, as Reclus argued that mistreatments of animals only increased in those human societies which were characterised by injustice and poverty: 'Between humans and animals, as between humans, justice can only result from friendship.'[54] In another letter, discussing the principles of socialism, Reclus wondered if the socialist idea could exclude animals, whose 'republican communities' he often mentioned as examples of mutual aid containing potential teaching for human communities: 'If we had to realise the happiness of all those who carry human figure and destine to death all our similar which have a muzzle and differ from us because they have a less opened facial angle, we won't realise our ideal. For what is on my behalf, I embrace also animals in my affection of socialist solidarity.'[55] From an ethical standpoint, Reclus did not find major differences between violence against humans and violence against non-human animals. 'I don't understand the murdering of an animal or of a human, I only do an exception for cases of personal or social defence.'[56] From a successive Reclus's letter, one can infer that Heath agreed on that principle: 'Like you, I give the same rights to animals and men, *and as the line must be drawn somewhere*, I try to trace it as far as possible, feeling perfectly that dog and cat are my brothers.'[57] Reclus summarised this principle in a letter to Heath's son, Carl, extending to the animals the right of resistance he discussed with Richard concerning Tolstoy. 'A far cry from Tolstoy, I believe in the possible use of force ... for the defence of the weakest. I see a cat tortured, a child beaten, a woman mistreated, and if I am strong enough to impede this, I will do.'[58] These discussions started in the 1880s between Heath and Reclus and then were extended to Kropotkin and Heath's British entourage: accordingly, the anarchist geographers played a role in the construction of the ideas which would inform Henry Salt's *Humane Review*, as I explain below.

Beyond friendship and philosophical conversations, business concerns and political ones likewise coexisted in the Reclus–Heath relationship. The British writer copy-edited Reclus's article 'Anarchy by an Anarchist' published by the *Contemporary Review* in 1884, which occasioned further discussions, revealing Reclus's confidence in a concrete outcome for his political wishes, and not least in a 'conversion' of his interlocutor:

> My friend, I really hope that before dying I will have the time to show historically that our anarchist ideas are not a simple dream. ... if we manage to publish the Letters on Anarchy from our friend Kropotkin [then jailed in France] ... I think that you will read them with pleasure and they will help you to modify your ideas.[59]

The conversation addressed the wider principles and problems of socialist schools. In England, Herbert Spencer was still a very influential author for activists committed to civil liberties, and Heath requested Reclus's opinion on this. Substantially matching Kropotkin's views,[60] Reclus praised the contribution which Spencer's ideas gave to early socialism but blamed his deliberate lack of an explicit socialistic perspective. Moreover, Spencer's lack of a 'humane' approach was seemingly pointed out by both Heath and Reclus, who wrote: 'Like you, I read Herbert Spencer's papers with a lot of interest, but like you I was shocked by many words and appreciations which does not seem humane to me.'[61] Humane concerns were connected to the principle of global and inter-specific solidarity, which the anarchist geographers substantiated in their critique of Malthusian theories, using their scientific authority to deny conclusions that others justified in the name of 'science', eventually the ineluctability of poverty for the 'natural' scarcity of resources. 'We want to extend this solidarity to everybody, as we know rationally, thanks to geography and statistics, that world resources are comfortably sufficient to feed everybody... It is in the name of science that we can tell the savant Malthus that he is wrong.'[62] Reclus's interest for critical analyses of authors such as Malthus and Spencer exposes how much the French anarchist geographer was directly involved with British and international debates.

Finally, anti-militaristic and anti-colonial concerns characterised the last exchanges between Reclus and Heath, in the period of the South-African war: 'In England, thousands and thousands of generous hearts bitterly suffer from injustices and atrocities deployed there and here against unlucky brothers ... and people with spirit of solidarity protest against the expeditions and raid massacres organised in so many distant countries. Noticing these abominable practices ... [we] shall remain united to give more strength to our resistance.'[63] The concept of the engaged scholar was summarised by Reclus as follows: 'I cannot wash my hands like Pilate'.[64] Heath proved to be sensitive to these arguments, as demonstrated by his contributions to the *Humane Review* that I discuss below.

In his very laudatory obituary of Reclus, equally published by the *Humane Review*, Heath tried for the last time to deny the evidence, arguing that Reclus 'was not anti-Christian'. However, Reclus might have not disliked Heath's statement that 'Jesus Christ would, I believe, regard him as a brother',[65] knowing the frequent use of the figure of Jesus by socialist propagandists at that time, albeit this was not intended to strengthen the Christian faith but to challenge the Church's influence on the popular masses.[66] It was in this spirit that Reclus wrote to Heath justifying these anarchists 'who stole the rich for giving to their brothers in Christ. Private property, here you are the theft.'[67]

Heath, and Kropotkin's British connection

Kropotkin's relations with Heath started in January 1886, on Kropotkin's return to London, when the Russian Prince appointed the British writer to copyedit his works, under Reclus's recommendation. The first documented contact is a letter from Heath on 15 January 1886, wishing Kropotkin 'many thanks for permitting me to read this pamphlet and coming article'.[68] Heath added queries in a matter of English phrasing and evoked a visit he received by the Kropotkins in former days. Therefore, Heath was one of the first to meet Kropotkin in London after his release from French prison. If we compare this body of correspondence with the contemporary exchanges between Kropotkin and Knowles,[69] it appears clear that securing Heath's collaboration was vital for Kropotkin in looking at the incoming editorial business for which *The Nineteenth Century*'s editor was soliciting him. Successive letters from Heath reveal details on the linguistic issues Kropotkin had to face during his first years in Britain, though they focused on idiomatic clarity rather than style: 'Revision consisted simply in using some words more familiar to ordinary readers than those you had employed, and in making the meaning of a sentence more apparent by a slight transposition of its parts.'[70] Heath noticed that Kropotkin employed frequent Gallicisms: his good command of French rendered him one of the best persons to do this work for Kropotkin in London.

Heath acted as an editorial agent not only for Kropotkin but also for his (anarchist) acquaintances. In 1887, he wrote to the Anarchist Prince that he had read 'the paper on Women in Old Russia, and I saw Mr. Bunting[71] yesterday about it. He promised to read it, and to write to M. Tchaikovsky about'.[72] Nikolai Vasilyevich Tchaikovsky (1850–1926) was an old friend of Kropotkin's, likewise exiled, and one of the promoters of the British petition for him. Heath, like many other British scholars, was involved in the publishing industry and acted as an editorial agent for refugees. This confirms the importance of sociability networks for understanding the construction of science, and the opportunities it offered to victims of political persecution. Apparently, this article was not published by Bunting, but the editor of the *Contemporary Review*, Knowles's competitor with whom Kropotkin had few direct relations,[73] was in touch with the anarchist geographers. In 1883, Bunting wrote to Reclus to request a paper, which was accordingly realised by Metchnikoff, on 'Evolution and Revolution'.[74] As mentioned on p. 161, Heath intervened in favour of Reclus with the *Contemporary Review* for 'Anarchy by an Anarchist'. In the 1890s, the *Contemporary Review* published Reclus's other articles, including 'East and West' (1894), which was considered one of Reclus's earliest attempts to question Eurocentrism in geographical representations.[75] It is arguable that Heath also played a role, in this case, in helping Reclus publish in English. In any case, the *Contemporary Review* looked sympathetically at the anarchist geographers, considering that, in 1894, it published a tribute to Kropotkin titled 'Our most distinguished refugee'.[76]

In his collaboration with Kropotkin, Heath also provided an introduction to specific aspects of English history and society; eventually, a list of references on agrarian statistics and the agrarian question as 'sources of information on the economical results of landlordism in England [which] can all be obtained at the British Museum Library'.[77] Some of the sources suggested by Heath, like the Parliamentary Reports, would be mentioned by Kropotkin in *Fields Factories and Workshops*. Again, analysing the fabric of the Anarchist geographers' works proves to be effective in tracing the origins of ideas and the importance of sociability networks for both resolving practical problems (eventually, English writing) and gathering ideas and information for intellectual work. As in Reclus's case, the conversation between Heath and Kropotkin soon took on a philosophical and political character, where some shared points on ethics could prevent ideological differences from stopping the dialogue. In 1887, Reclus reassured a concerned Heath about Kropotkin's temporary silence. 'You don't know our friend Kropotkin if you think that he could be angry with you because you disagree with him on some points in moral or history.'[78] Reclus compared explicitly these two parallel exchanges of correspondence giving a synthesis valid for both. 'We were often in disagreement, what does not impede mutual respect and affection.'[79]

As in Reclus's case, Heath often requested Kropotkin's opinions on the political concerns of the day, asking for instance: 'What do you think of Gladstone? Did you see what Clemenceau said about him?'.[80] As Reclus observed, Heath shared with the socialists a common commitment to 'the cause of justice'.[81] It is highly unlikely that Kropotkin could write to Heath praising Gladstone, and *Freedom*'s positions on these matters were discussed in Chapter 4, but it is significant that Heath introduced Kropotkin to British debates by expressing strong support for Gladstone's project for Home Rule, which was under discussion at the Parliament in those weeks. According to Heath, 'This Home Rule question means more than at present appears. It will, I believe, presently develop into a struggle between the oligarchs and the democracy of England.'[82] Kropotkin's correspondent understood that 'the whole body of the aristocracy [stands] against Mr. Gladstone who on the other hand is universally applauded by the ... people'.[83] If some of Heath's definitions—such as that of the 'Homeric hero' for Gladstone—were likely to sound ridiculous to an anarchist's ears, Kropotkin's availability for these conversations was not only due to him pleasing a professional acquaintance, but also to his interest in listening to and understanding the most progressive parts of British society and scholarship, to which he was mostly speaking through works such as his *Nineteenth Century* papers.

Meanwhile, Heath proposed to the Anarchist Prince conversations with 'young friends [willing] of hearing apostles of a true form of socialism ... a sympathetic and intelligent audience'.[84] The first summer that the Kropotkins spent in Britain included visits to Eastbound and the East Cliff under Heath's advice.[85] More importantly, it was Heath who put the Russian exile in contact with the Fellowship of the New Life. One of the founders of the Fabian Society and member of the Fellowship, Percival Chubb (1860–1960), had asked Heath to engage Kropotkin 'to give a lecture'[86] when coming back to London.

As with Reclus, Heath discussed anarchism with Kropotkin, giving uncharitable definitions of this term as something very far from his ideas, and even fearing that it might serve as a possible justification for 'terrorism'.[87] His ideas might partially match Kropotkin's in blaming the so-called 'illegalist' tendencies, not rare among French 'individualist anarchists', often justifying theft and murder as instruments of struggle against society. Eventually, Heath evoked the recent trial of Clément Duval (1850–1935), a Frenchman who claimed the 'anarchist' nature of his activities as a thief, arguing that 'he did the very opposite to what anarchist socialism set before as the moral ideas of life'.[88] *Freedom* defended Duval against the disproportionate repression of the French state, which sentenced him to death and then to deportation in Cayenne. In general, social and communist anarchists expressed solidarity for people like Duval, a poor proletarian compelled by misery to steal, but blamed the example that they set. It was the case with Jean Grave, who wrote to Duval, years later: 'I defended you for your sincerity, but you gave a very bad example because, after your trial, all the ruffians invaded our movement claiming your example to justify their greed.'[89] Certainly a humanistic inspiration characterised communist anarchists in their struggle against individualism and Stirnerite tendencies.[90]

Heath's biography confirms that the British writer believed in a sort of Christian socialism inspired by radical reformation movements such as the Anabaptism, which he considered to be 'not dead; it slumbers in the hearth of the poor man, and will assuredly rise again',[91] defining men such as Mazzini, and then Reclus and Kropotkin, as the modern prophets of 'Liberty, Equality, Fraternity'.[92] Heath expressed his idea of Christian socialism in a letter to Kropotkin: 'What lessons in Republicanism the Huguenots at Nantes, the Calvinists in Switzerland and Holland and the Puritans in England and the United States gave at various times. Louis Blanc recognises that the true genesis of the Revolution was a Christian one.'[93] In copyediting Kropotkin's autobiography, Heath stated that '[your] account of the English socialist movement is most inadequate',[94] as it did not consider Christian movements. Something which Heath could not understand in Kropotkin's optimism was 'by what means [was he] going to eradicate the natural selfishness and egoism of the individual man'.[95] Despite some common concerns in ethics, Christianity and Anarchism remained antagonistic and mutually incompatible concepts in these discussions.

As in Reclus's case, Heath and Kropotkin debated a varied range of topics. In 1889, they started discussing the French Revolution, exchanging books[96] and ideas on the revolutionary terror;[97] after that Kropotkin was requested by Knowles to write the papers which inaugurated his series on the *Great French Revolution*. Proofreading *Fields, Factories and Workshops* allowed Heath to provide information and suggestions on the differences between 'French and English rural economies'.[98] Discussing the idea of mutual aid, Heath seemed convinced by Kropotkin's argument on the need for cooperation as a basis for human ethics, and also by his defence of so-called 'savages'. Instead of considering them as 'backward', Kropotkin deemed 'primitive' societies as examples of mutual aid and application of moral feelings by human communities striving together against environmental challenges. Health concluded: 'Dear friend, you are doing good work',[99] and extended the concept of Mutual Aid to his works on the Anabaptists and English peasants. What divided them, like in Reclus's case, was the anarchist discussion on the possible uses of violence, though in both cases it seems that the problem was more in the definitions than in the substance. After reading the French edition of Kropotkin's *Conquest of Bread*, Heath observed: 'In reading your book, *La conquête du pain*, so many thoughts have come and gone that I should have liked ... to have told you ... a violent revolution cannot in itself alter human nature, but may even stop and prevent the natural evolution of things.'[100] Heath even used harsh terminology to criticise anarchism, declaring himself 'repelled'[101] by this definition and expressing pessimistic thoughts on the fact that 'one cannot possibly reform this world wholesale'.[102] This led Heath to ask amazing questions, at least from the standpoint of his interlocutors: 'Men of such wonderful mental form and universal knowledge as yourself and Élisée Reclus [represent anarchism], and I am propelled to ask myself what is the force which is driving you to the champions and apostles of Anarchy.'[103] The contradiction was clear: all along his life, Heath struggled in making sense of his rejection of anarchism and atheism while admiring and personally liking their main 'apostles'.

Heath was especially interested in all that circulated in British humanitarian and humanistic groups, sometimes characterised by forms of mysticism. In 1891, he signalled to Kropotkin the rise of theosophy, a philosophy to which former *Freedom* supporters like the aforementioned Annie Besant adhered. 'The new theosophy movement affords material for thought. Especially as a sign of the times.'[104] William Morris was included in the circle of

Heath's authors of reference, as he declared after Morris's death. 'Socialism has lost another notable champion and leader in William Morris. I believe that he would have had very little sympathy with my way of looking at things and perhaps would have summed me up even more severely than he did Gladstone, but for all that I admire him ... he had a deeply poetic, romantic mind and loved justice with an intensity'.[105] In the same letter, Heath detailed a visit by Élisée Reclus. Amazingly, in the following year, a false rumour about Élisée's death was the occasion of a premature tribute to the French anarchist geographer, for which Heath wrote: 'How much I owe to him! Often have I said it was he who inspired me with a love of knowledge, with a real interest in life. It was a kind of revelation my coming under his instruction.'[106] Eventually, Heath expressed his affection for other figures of the Reclus family, that is Louise Dumesnil and Paul Reclus. One can imagine Kropotkin's astonishment in receiving this letter with such unexpected (and fake) news!

During the years of the South-African War, one of the principal matters of discussion between the two men was the problem of jingoism and, in this case, Heath's views well matched Kropotkin's opposition to war and colonisation. Like the *Freedom* group, Heath attributed the popularity of pro-war positions to the weakness of the worker's movement. 'Members of parliament, judges, professors, generals etc., all enthusiastically support it in the person of Rhodes ... Again I cannot but see that the forces to oppose this new reactionary movement are sadly feeble – witness the whole field of the socialist movement in England.'[107] Likewise, Heath was scandalised by the behaviour of most socialists, including the Fabian leaders mentioned in Chapter 4. 'I find to my dismay leading Fabians may be imperialist and talk of "niggers" and the inferior races.'[108] The question of racism was part of this critique, which not only focussed on the opposition to one war but on wider pacifist agendas. According to Heath: 'This is not only a war between the British and the Boers. It is British Imperialism against Humanity.'[109]

The opposition between the idea of humanity and imperialism was confirmed by Heath in his contribution to the first number of the *Humane Review*, in 1900, titled 'The Kafir and his master'. In this paper, Heath did not distinguish between British and Dutch colonists in South Africa and fumed at the 'burden' that the 'white man' brought there. 'In a black man's country, he regards the natives with contempt and suspicion and continually encroaches on their land and their liberty. What is most ominous is that his aversion to them increases, and the gulf is wider than it was a century ago.'[110] Rejection of racial inequality was a key point in the analyses of Heath, who also stressed the possible inversion of the relations of force due to demographic dynamics, and condemned Rhodes for the massacre of the Matabele. Ironically, Heath appears equally as one of the forerunners of the aforementioned 'Rhodes Must Fall' campaign, addressing the concerning popularity of Rhodes among university youth in an 1899 letter to Kropotkin: 'What do you think of the welcome to Rhodes at Oxford? The tremendous manifestation in his favour by the *Jeunesse Dorée* of England. What will they not forgive when a man goes in for their ends and objects'.[111] Heath was referring here to the former support Rhodes had given to the Home Rule, blaming his opportunism, but what stands out is the networking of British-based intellectuals who, at the top of Rhodes' popularity, already considered this person as a symbol of imperialism, militarism, and colonial brutality. In short, Heath's denunciations of colonial crimes matched the essential of Reclus's and Kropotkin's positions on these topics.

Discussing the Dreyfus affair in France, Heath considered it as a 'Little Transvaal', that is an expression of militarism and chauvinism. 'How disgracefully the same military spirit of brutal cruelty is acting in France. Dreyfus reconvicted and sentenced to 10 years of imprisonment!'.[112] More extensively, Heath denounced the pervasive nature of imperialism in British culture and society, beyond the responsibility of the politicians, a problem stressed by Kropotkin and by the anarchist press. 'The Imperialistic idea has been carefully washed up, as in the Imperial Institute etc. Also the press in London has been bought.'[113] Heath's outrage for the fact that a journalist could develop some arguments for money and not according to conscience clearly owed to his puritan inspiration, but at the same time it resonated with anarchist arguments considering the individual's ethics as more important than laws and economics or personal interest, as it was the case with anarchist calls for desertion and objection of conscience.[114] Again, the difference was Heath's religious inspiration, as he explicitly put his hopes in a divine kingdom more important than the British Empire, while for the anarchists the alternative society should be first and foremost human and humane.

The occasion for Reclus's commemorations, which occupied several of Heath's letters, came at Reclus's death in July 1905. The English historian wrote to Kropotkin: 'The world has lost a great man, and me a most faithful friend ... I have known him perhaps as long as anybody outside his family circle and friends, but ... from those early days ... when he gave me some lessons in French to recent times, I have had to know him mainly though his letters and his rare visits'.[115] It is true that Heath was one of Élisée Reclus's longest acquaintances, as they were in touch, though with long interruptions, from 1852 to 1905, and one of his more prolific correspondents, judging only from Élisée's responses. To express his admiration for his lost friend, Heath used a geographical metaphor: 'As Mount Blanc among the Alps, Élisée always appeared immeasurably higher than other men. In his later days, he seemed to grow sweeter and even tender, just as you might expect Mt. Blanc to appear at sunset.'[116] Kropotkin had requested from Heath biographical information on Reclus's early days for his obituary. Although he did not keep any notes of that period, Heath was willing to offer his recollections, which however did not add anything ground-breaking to the biographical information that Reclus's letters furnish on his early British period. It is worth noting that Heath had direct relations with multiple members of the Reclus family and their French acquaintances. Some years earlier, he had informed

Kropotkin that: 'Élie and Mme Élie were there, also the Noël[117] family and Regnier and Paul Reclus came for a day or so'.[118] Also, Heath's sons were in contact with the Reclus family and with the Dutch anarchist Ferdinand Domela Nieuwenhuis, a close friend of Reclus.[119]

After reading Kropotkin's obituary for Reclus, Heath praised this work arguing that the Russian was the best-placed person to write such, 'having a profound knowledge of his thought, both in geographical and social science and such an intimate experience of his character'.[120] Heath's letter highlighted both Reclus's and Kropotkin's insertion in the British 'humanitarian' circuits: *The Humane Review* had published Reclus's paper 'On Vegetarianism' (1900) and now Heath wrote that he accepted 'our friend Henry Salt's invitation to do [Reclus's obituary] in the *Humane Review*'[121] requesting, at the same time, Kropotkin to be for once the reader and copyeditor of this paper, inverting the traditional roles. If Heath's focus was unsurprisingly on the Protestant background of Reclus's family, common grounds on equality, justice and 'humane' consciousness emerged between Richard Heath and the anarchist geographers. Humanism was a topic directly addressed in Heath's letters. While copyediting some of Kropotkin's papers, he suggested the Anarchist Prince assume 'a still more evidently humanistic vein',[122] including a shared commitment to education with 'manual training'.[123]

From the end of the nineteenth century, new authors and concepts revived the two men's discussions of violence. Heath tried to introduce Friedrich Nietzsche to Kropotkin, and this was highly ironic because, among a certain number of (self-declared) anarchists, Nietzsche's influence was associated to that of Max Stirner to justify indiscriminate violence, a practice which Kropotkin and Reclus strongly condemned. Nonetheless, for Heath:

> I do wish you would grapple with Nietzsche. He seems to me to have got to the heart of the truth and to try to squash the very source of the revolutionary movement in modern Europe. I could only get you to see it as he sees it in his Antichrist I should not have lived in vain. The Christianity on which he poses out his work and contempt is exactly the True Christianity in which I believe.[124]

In the following letters, Heath insisted that Kropotkin must read Nietzsche. Kropotkin had read Nietzsche's works but did not appreciate them, considering the German author as 'a philosopher in carpet slippers [and] the first Philistine'.[125] However, the author whose work, well known by Kropotkin, stimulated the debate was again Tolstoy, considered by Heath as one of his main inspirations together with 'a Victor Hugo, a Mazzini, a Lamennais or a Lincoln'.[126] Heath highly valued Tolstoy as the man who represented 'to our age what Rousseau was to the eighteenth century'.[127] Discussing with Kropotkin the evils of the repression which followed the 1905 revolution in Russia, Heath praised 'Tolstoy's courageous protest against the Tzar and his government. It was the word of a Prophet.'[128] Heath indicated Tolstoy's

Christian and non-violent anarchism as the alternative to what he called the 'terrorism' of the former generation of Russian revolutionaries. He also informed Kropotkin of his opinion about Reclus's remarks on the Blaricum commune mentioned above: 'I had a letter ... from Élisée ... about the anarchist communist colonies in Holland and Belgium. ... he thinks they would have done better in defending themselves than in going elsewhere. I am not at all sure of that.'[129]

Heath's later letters increasingly dealt with mystic issues in a way that might have become more and more difficult to follow for Kropotkin. The last dated letter was sent on 4 September 1910. However, his correspondence of almost 25 years filled the largest folder of letters received by Kropotkin from a British correspondent, together with Keltie's folder. Like Keltie, Heath played a role in Kropotkin's insertion in British circuits and, to a minor extent, he did the same for Reclus. In both cases, personal feelings and sociability networks intersected with political conversations and especially with editorial and professional endeavours. It was the case with Heath's works (copyediting Reclus's and Kropotkin's papers in English) and with Reclus's facilitation of Heath's work in France concerning Quinet and other topics that interested him on the Continent. Ironically, Heath had more problems than his famous anarchist friends in entering the publishing field and gaining public success: in his letters to Kropotkin, he often complained that he hardly reached journals like *The Nineteenth Century*. It was the case with a paper criticising institutional churches from a theological standpoint, which Heath had sent 'to Mr. Knowles [who] refused to consider it'.[130] Finally, if Reclus and Kropotkin were both strongly opposed to Heath's views on a matter of faith and religion, they agreed in the principle of human[itarian]ism. I do not insist further on the Protestant roots of the Reclus family because the anarchist geographer publicly repudiated religion and embraced atheism, but it is clear that some common points between his anarchism and traditions of radical and dissenting Protestantism played a role in his friendship with Heath and other British acquaintances.

Heath and Reclus both collaborated with Salt's reviews, as Heath and Kropotkin discussed: 'Have you seen M. Salt's new *Humane Review*?'.[131] Kropotkin did not contribute directly to the *Humane Review*, but his 'Humane Science' lecture was published in *Humanity*, as I explain below. However, before introducing the *Humane Review*, it is necessary to discuss the works of a couple of Reclus's and Kropotkin's important British friends, Edward Carpenter and Havelock Ellis.

Anarchism, humanism and gay rights: Edward Carpenter and Havelock Ellis

As with activists akin to the Fellowship of the New Life, Edward Carpenter was a figure haughtily discarded by Marxist historiography. For instance, E.P. Thompson claimed that 'Carpenter's revolt was individualistic, undisciplined, and (backed by a legacy of £6,000) not especially arduous'.[132] In his famous book on Morris, the 'Big Man' of British Marxist historiography also ignored Carpenter's commitment to homosexual rights: perhaps, Thompson considered this topic as excessively '*petit-bourgeois*', or not worthy enough for his idea of revolution. In 1977, in their monograph on Carpenter and Havelock Ellis, Sheila Rowbotham and Jeffrey Weeks argued that these two authors were associated by the fact they were both 'dismissed easily as cranks and visionaries',[133] despite the fact that their respective works anticipated later views of humanitarianism, sex, love, and civil rights by challenging late Victorian and Edwardian cultural prejudices, especially 'the authority of the father and husband or of an ideology which stressed wifely submission, rejected sexual pleasure and preached salvation through abstinence and thrift'.[134] Both early Fabians and members of the Fellowship of the New Life, Carpenter and Ellis diverged in their respective purposes. 'Ellis was interested in developing an understanding of sexual psychology, while Carpenter was more of a populariser of sexual theory.'[135] To well understand the political context of Reclus, Kropotkin and the *Freedom* group's engagement with these topics, it is worth considering Rowbotham and Week's remarks on the contemporary attitude of Marxists.

While Marx and Engels maintained a 'bias towards heterosexuality' by defining sodomy as an 'abominable practice',[136] the anarchists of *Freedom* supported Carpenter, Ellis and the activists of the Legitimation League and intervened in the famous 1895 case of Oscar Wilde, likewise an acquaintance of Kropotkin, as they had met in William Morris's house.[137] As I describe below, *Freedom* blamed Wilde's condemnation for what was then called 'gross indecency' or 'buggery', albeit taking all the phrasing and terminological warnings necessary in a period when it was impossible to take a public position in favour of homosexuality without being arrested. The main argument of this section is that the association between anarchism, civil liberties, and sexual liberation was possible thanks to anarchist commitment to the cause of 'humankind' rather than only class struggle, and it was directly and indirectly favoured by the intellectual networks of Kropotkin and the Recluses. This challenges commonplace assumptions on the supposed 'endemic—if nowadays more carefully hidden—homophobia of the left and the labour movement'[138] showing early anarchist engagement on that front, yet paralleled by other early anarchists mentioned in David Berry's paper on the dialectic of homosexuality and revolution, like the French individualist Emile Armand (1872–1963).

Several works on Edward Carpenter mention his association with Kropotkin, observing that, like Wilson, Kropotkin was invited by Carpenter to give socialist lectures in Sheffield in 1887.[139] The region of Sheffield, where Carpenter was based, was the object of some of Kropotkin's pages on the 'petty trades' of big industries addressed in *Fields, Factories and Workshops* and mentioned above, thanks to Carpenter's commitment as a local informer.[140] Indeed, the letters surviving in Moscow confirm that Carpenter sent Kropotkin notes on Sheffield manufactures in 1888. They also reveal the early acquaintance between the two men when Carpenter wrote: 'I shall be glad to get a little talk with you again next time'[141] on the occasion of Carpenter's visits to London. Carpenter wrote several letters to Kropotkin insisting on securing his lectures for the local Socialist Society, where the Anarchist Prince was welcomed to talk on a subject chosen among: 'The state of Russia, the state of Europe, or the general programme of anarchism.'[142] However, he was warned that: 'Our audience is mostly poorly educated, and able to appreciate special illustrations better than general laws ... The special doctrine of Anarchism is hardly known'.[143] This was a classical problem of popular anarchist propaganda: knowing anarchism requested a minimum of culture and literacy which working classes did not always possess. This explains why the efforts of intellectual propagandists and especially of the intellectuals such as Carpenter, Kropotkin, and Reclus were directed continuously towards popular audiences. This was also in line with the anarchist principle of individual consciousness: only if all the members of a social group develop critical and intellectual skills at the individual level, would the decision making be libertarian and egalitarian in the collective.

Like Dryhurst, Carpenter was one of Kropotkin's 'Western' friends who remained in contact with him until his later years, and who declared a special 'heartily affection for him',[144] as Carpenter wrote to Kropotkin's wife Sofia (whom he addressed as 'Dear Princess') in 1905. In 1916, Carpenter announced a visit to the Kropotkins, who lived then in Brighton[145] and, in 1917, he expressed his joy for the February Revolution in Russia, arguing that this outcome was the result of 'solid work that can never be undone',[146] and rejoicing in the fact that the elderly Kropotkin finally had this satisfaction after so many years of political delusions.

As his letters to Marsh quoted below also show, Carpenter was a collaborator and a supporter of *Freedom*. In December 1892, the anarchist journal published a paper entitled 'Important letter from Edward Carpenter'. In this text, Carpenter intervened in the ongoing commemoration of the Chicago martyrs and of the Walsall anarchist, arrested under explosive charges but in fact instigated by an agent provocateur, a case which directly affected Carpenter's acquaintances in Sheffield.[147] Carpenter addressed there the problem of violence in the anarchist debate; despite the fact that he deemed it absurd to deny that the option of violence should be contemplated, he argued that in both Chicago and Walsall cases, 'we have men of the most gentle, humane and peaceable disposition', including someone who 'would not hurt a fly if he could help it'.[148] If it is significant that Carpenter used here the category of 'humanity' in association with the respect of animals by mentioning the fly, what followed was a classical anarchist argument to defend violent acts accomplished by proletarians pushed to do so by their exasperation for living in an unjust society, without availing the principle of violence itself. 'What more serious indictment of existing society and institutions could we have than this? When society drives its best men to such extremity, how rotten indeed must it be?'[149] Finally, Carpenter fumed at the 'barbarous' retaliation operated by bourgeois society in condemning the five Chicago anarchists to death and the four Walsall ones to heavy detention penalties. Regardless of the validity of such arguments, it is worth noting that Carpenter spoke here not as a guest, but as a stable collaborator of *Freedom*. Carpenter's biographers suggest that he oscillated between different political groups and his best definition was perhaps in his *Freedom* obituary in 1930, stating that: 'More Anarchist than Socialist ... he never cared to label himself'.[150] However, the papers and correspondences I analysed allow considering him as a representative member of British anarchism for at least all of the 1890s, and doubtlessly for many of the following years. The correspondences Carpenter exchanged with Marsh, surviving at the IISH, show Carpenter's militancy in favour of *Freedom*. In 1899, Carpenter requested Marsh to send him 'any circulars or leaflets printed'[151] for distribution, concurrently subscribing to the journal for £ 3. In 1900, Carpenter encouraged Marsh to continue campaigning against the colonial war in South Africa: 'In these days of Government megalomania, it is so good that you still continue to protest'.[152] It was in his capacity of renowned anarchist exponent that Carpenter intervened in the Wilde case from the columns of *Freedom* in 1895.

In Victorian England, 'buggery' was a crime punishable with 'up to two years' hard labour'.[153] The famous writer Oscar Wilde had incurred this charge, for which he was sentenced and imprisoned in 1895. *Freedom* took to his defence in June, arguing that 'all persons except the hypocrites ... must see that no one principle of truly human morality has been vindicated'.[154] It is worth noting that explicit stances in favour of homosexuality could hardly be printed at that time, if one wished to avoid being sentenced for press crimes. For that, the article published by Carpenter in July looked very prudent in its terms and circumlocutions, but not less clear in the essential, that is the disapproval of a society so 'inhuman' to punish 'deviant' feelings and sexual behaviours. In the Wilde case, Carpenter first lamented the ferocious attitude of a substantial part of the public opinion. Then, he proposed a reflection on the general censorship existing on sexual matters, including in the socialist field, which he defined as: 'The general tendency to relegate all these matters to a kind of ominous silence, and to refuse absolutely to discuss them'.[155] For Carpenter, this was not only a political issue, but first a social one:

> Is it not largely to the rather cowardly silence on this particular subject that we must trace the state of our schools? And is it not a shame, simply on this account, to leave young folk to get themselves into dire trouble and confusion of mind, without a word of help or guidance to them? Owing to this state of affairs there is undoubtedly in the public mind almost complete ignorance of some facts of human nature, which are really very important.[156]

If this plea for sexual education and for considering the importance of these topics for human life was undoubtedly a challenge to Victorian morality,

Carpenter's following attempt to pose the question of the existence of different sexual behaviours, associating them with 'love' and not with 'vice', was likewise courageous.

> It appears that, a considerable percentage—perhaps one in every fifty—of human beings are so born that they can only love (using the word in its best and hugest sense) others of their own sex ... In some cases, for social and other reasons, they marry, and may have children; but even so, they do not, and apparently cannot, really love their partner, and the marriage is generally unhappy for both parties.[157]

For Carpenter, silencing and hiding these phenomena was the worst way to deal with them for society, and a great cause of trouble for both girls and boys coming to their sexual maturation and feeling inclinations for what the author defined as a 'homogenic temperament'.

Struggling to find a term for defining these persons as a social group (to which he belonged), Carpenter used the definition of class, which might seem odd today but which expressed well the sense of his final claim: 'Surely, in face of the sufferings of this class and the difficulty and reality of the problems which they have to face, it is time that there should be some sane and impartial consideration of the whole subject.'[158] Again, it was not an outsider speaking, but a leading figure of the anarchist movement, who authored a paper on the first page of the leading journal of anarchists in Britain. I insist on the importance of texts and contexts because, regarding famous figures like Kropotkin, one finds often hasty judgements which lack serious ground: for instance, in a book on anarchism and homosexuality in the United States, Terence Kissack, after deeming the jail term served by Kropotkin in France 'short' (it was more than three years), attributes to the Anarchist Prince a 'harsh condemnation of homosexuality'[159] in the book *Russian and French Prisons* (1887), where homosexuality was never mentioned! Kissack's judgement was based on the suggestion that the remark by Kropotkin on 'breeches of moral law', with mention of other general works on prisons, referred to homosexuality without providing evidence for this suggestion, or at least conceding that Kropotkin might have changed his mind by fraternising with comrades like Goldman, Ellis, and Carpenter, as is often described in Kissack's invaluable work. *Freedom*'s story shows that, in opposition to other socialist schools more greatly influenced by Marxism and economicism, the anarchists were more likely to support early campaigns for lesbian and gay rights due to their consideration of all humankind, and not only of a certain class, as the beneficiary of full social emancipation.

In 1897, Carpenter launched the concept of 'humane sciences', organising a series of *Humane Science Lectures*. Kropotkin contributed to this series of conferences, whose contents were subsequently collected in a volume, whose last chapter was the text of Kropotkin's lecture, 'Natural selection and mutual aid', which I reproduce in Appendix A. What is important is that this very little-known text was not written directly by Kropotkin, but reconstructed by Carpenter from attendants' notes, and concurrently published in the journal *Humanity*, summarising the humanistic side of Kropotkin's ideas on evolution. Carpenter aimed to put science closer to human beings, their feelings, their daily lives and needs, conceiving the practice of 'science', in its broadest sense, as an exercise targeting human betterment and fraternity. The book was an original and neglected publication which anticipated contemporary claims against the 'objectivity' and 'authority of science', proposing relativistic views of the ideas of civilisation and linear progress. Carpenter's endeavour was declaredly a protest against what he defined as 'the attempt of Model in Science to get rid of human feeling and to look at everything in the dry light of the intellect'.[160] In a time of strong positivist hegemony, these claims anticipated several ideas of so-called humanistic geographers who reacted against the hegemony of quantitative approaches between the 1960s and the 1980s.[161] At that time, geographers such as Anne Buttimer, Yi-Fu Tuan, David Seamon, and David Ley criticized what they called 'the *danse macabre* of materialistically motivated robots'[162] which they considered to have exerted a hegemony on geography, 'for the impenetrable appearance of objectivity to be demystified to illuminate the shifting anthropocentric presuppositions masked by the firm categories and rigorous procedures'.[163] Significantly, in different times and places, 'humane sciences' and 'humanistic geographies' challenged assumptions on objectivity, ethnocentrism and anthropocentrism.

In his lecture, Carpenter argued that 'many of the laws of Science enounced as universal truths, were of insufficient application only, that many of the conclusions, so strongly insisted on, were of quite doubtful validity; and at last this increasing dissatisfaction culminated in a rather violent attack or criticism of Modern Science that I wrote and published about the year 1884'.[164] Likewise, Carpenter argued that, separating 'the intellectual in man from merely perceptive, the emotional, the moral ... this modern science was leading to a narrow-mindedness and dogmatism as bad as the old'.[165] Carpenter was referring to religions: his main argument was that modern science pretended to get rid of dogmatism and superstition, but instead it reproduced dogmatic thinking and took the place of religion without giving up the principle of authority. This entailed that scientists considered to be legitimised in determining what universal knowledge, that is 'truth', should be. Carpenter matched Kropotkin's criticisms of the 'new religion' built by positivist scholars like Auguste Comte.[166] Carpenter also mentioned Reclus's works, published in French, on nature, culture, and civilization, which he later translated for Salt's *Humane Review*. Meanwhile, he developed Reclus's argument for the 'free alliance for various purposes of the primitive man with animals'[167] to be intended not only as evidence supporting interspecific ideas of mutual aid but also as an important insight for the so-called 'civilised'. For Carpenter, who published his criticisms of the idea of civilisation, likewise drawing upon Reclus's and Kropotkin's works:[168] 'It is notorious that in many respects the perceptions, the Nature-intuitions, of savage races far outdo those of civilised'.[169] The example of the 'primitives' was used to value different kinds of knowledge, based on feelings and daily experiences rather than mere rationality, for which Carpenter concluded that, 'in Natural History and Botany ... we have hitherto not only neglected the perceptive side, but also what may be called the intuitive and emotional aspects'.[170] The following step was questioning anthropocentrism by comparing this kind of knowledge to animal's instinct: 'Primitive peoples have a remarkable instinct of the medicinal and dietetic uses of herbs and plants, an instinct which we also find well developed among animals—and I believe that this kind of knowledge would grow largely if, so to speak, it were given a chance'.[171] If comparisons between 'primitives' and animals were not unusual in scientific literature of imperial ages and the construction of scientific racism, in the case of Carpenter and Reclus, the terms of the discussion were completely inverted, because they considered the skills of animals and of different human groups in adapting to a certain environment as a key example in developing cooperative capacities in evolution. The target of education, according to Reclus, would have been 'the plenary union of the savage with the civilised and with nature',[172] which meant that not only ideas of 'civilised' superiority had to be questioned, but also the dissociation between nature and culture, world and mind.

The second 'humane lecture', by Scottish naturalist and Geddes's friend John Arthur Thomson (1861–1933), drafted a 'humane study of natural history', which was intended to be an approach respectful of non-human protagonists of that history. Thomson especially complained against the extermination of animals for eating them, for feather and other commodities, making cases for early protectionism and vegetarianism. For first peoples, as well as for non-human (animal and vegetal) communities, the evil of civilisation was always extermination. 'One of the sad biological facts of to-day is the extermination of many beautiful and noble forms of life. We probably breed new species of Bacteria, but we exterminate bisons and beavers, and how many more'.[173] In the first volume of *L'Homme et la Terre*, Reclus considered early human imitations of animals as one of the first drives towards human material progress in pre-historic human communities, defining animals as 'close brothers' to humans, 'animals themselves'.[174] Thomson quoted Reclus on his idea of human and more-than-human 'kinship': 'The herbs and the trees, the birds and the beasts, sent their tendrils into the human heart, claiming and finding kinship'.[175] Also for Thomson, the basis of knowledge was not limited to rationality, but lay in the 'unity of life [i.e.] the trinity of knowing, feeling and doing, of brain, health and hand'.[176] Among the other contributions to the book edited by Carpenter (in addition to Kropotkin's lecture reproduced in Appendix A), a chapter by W. Douglas Morrison discussed prisons, while the physician and hypnotist John Milne Bramwell questioned the scientific authority of medicine.

Carpenter's archives in Sheffield show that, in his late years, he corresponded with Paul Reclus, who requested his collaboration for the *Temps Nouveaux*, [177] and remained in touch with transnational anarchist networks. In his autobiography, he recounted how he managed to help Japanese anarchist Sanshirō Ishikawa, a great admirer of Élisée Reclus, sheltering him and putting him in touch with Paul Reclus in Brussels. 'We tried to find him employment and a means of living ... but without success, and after similar efforts in London he migrated to Brussels where he knew of a friend in Paul Reclus, son of Élie and nephew of Élisée Reclus, and where he obtained an occupation in decorative painting. This was early in 1914'.[178] In that same year, the relevance of Carpenter's figure for the anarchist movement was shown by the fact that *Freedom* celebrated his 70th birthday with a warmly congratulating paper, a tribute which had been only paid to Kropotkin two years earlier when it was Carpenter who gathered recollections of Kropotkin's fellows and 'friends in Great Britain and Ireland'[179], printing an address which I reproduce in Appendix C.

Havelock Ellis was Carpenter's friend and, although he was more committed to intellectual criticism of Victorian society rather than political activism, he shared the essential points of Carpenter's engagement. Likewise sympathetic with feminism, Ellis is considered by his biographers as a protagonist of the traditions of British socialism which promoted a 'humanist ethical revolt against capitalism'.[180] Committed to evolutionism, Ellis published studies on *The Psychology of Sex* whose part on homosexuality, *Sexual Inversion*, was considered as a plea for homosexual rights, albeit remaining upon the level of a scholarly discussion. Significantly, the police identified this 'lewd, wicked, bawdy, scandalous libel' with anarchism, as such work would have certainly provided 'a convenient hammer with which to crush the society'.[181] Indeed, in 1898, the editors of *Freedom* showed solidarity with George Bedborough, the secretary of the Legitimation League, an association initiated by anarchists which supported the rights of 'illegitimate children' and free love. An early advocate of vegetarianism like Carpenter, Bedborough was eventually arrested for selling copies of Ellis's book (although the author was never affected by repression at that level) as well as other publications of the League. This excerpt from a *Freedom* article is indicative of the prudence which addressing these topics requested.

While the case is being judicated upon it would not be wise to say what we think about it; but we shall say what we intend to do about it. First, let every reader of *Freedom*, send on any little sum he can spare to the defence fund. Of course, we are all poor and are continually giving, but here is a man attacked because he wishes to have the sex question discussed in all its phases. If it was one of us connected with *Freedom* I don't think we should trouble the legal profession for our defence ... But we also believe in liberty for others and let us help them to defend themselves as they think best. Secondly, on this question of Sex, *Freedom* has always taken the position that, for the mass of the workpeople, until the economic conditions were righted there could be neither freer men nor free women; and, while never afraid to discuss it fully, have not thought it wise to give undue space to it. For our part, we think we understand the relative importance of this subject; know that sex-relations are part and parcel of and have grown out of, property conditions; but at the same time we know that some, whose economic conditions do not trouble them, may investigate the sex question from a scientific point of view and will by that study be led on to study other parts of the social problem. The love of liberty is opposed to statute law and ignorant conventionality in any direction, whether conscious or not, certainly makes for Anarchism.[182]

At the end, Bedborough pleaded guilty and was sentenced to a mild term, but the League did not survive the shock. At least, this was the occasion to create a Free Press Defence committee including Carpenter and socialist leaders like Shaw and Hyndman.

In February 1913, *Freedom* published a review of Ellis's and Theodore Schroeder's *Witchcraft and Obscenity*, noticing that this book 'draws a parallel between belief in witchcraft and belief in obscenity. When people believed in witches, they burnt the witches; just as our own courts to-day join with the obscenity-hunters to affirm that obscenity is in a book and not in the reading mind, and that therefore the publisher, and not the reader, shall go to gaol for being "obscene."'[183] This way of pointing out the relativity of common beliefs was in line with intellectual anarchism, and with ideas of evolution from superstitions and dogmatic knowledge to the free and secular exercise of critical thinking.

> When men ceased to believe in witches, witches ceased to be; and so when men shall cease to believe in the 'obscene' they will also cease to find that. Obscenity and witches exist only in the minds and emotions of those who believe in them, and neither dogmatic judicial dictum nor righteous vituperation can ever give to either of them any objective existence.[184]

Civil rights and the cause of free speech were then strictly linked in the scholarly and intellectual milieus discussed here.

Significantly, Ellis was acquainted with the Recluses and interested in their works, especially Élie's. In his recollections, Ellis recounts a period when he settled for three months in Paris, in 1890, to improve his studies of medicine. Patrick Geddes had furnished him the contacts for settling and for contacting Parisian scholars. Ellis went to the Paris Anthropological Society where he met Paul Topinard (1830–1911), an acquaintance of the Recluses despite his relatively conservative positions, and Léonce Manouvrier (1850–1927), a progressive anthropologist,[185] correspondent of Jean Grave and broad supporter of anarchism.[186] Another member of the Society, Charles Letourneau (1831–1902), was likewise praised by Ellis as 'a champion of free thought'.[187] But the figure who most impressed Ellis in Paris was the elder of the Recluses, Élie, then director of the Hachette's library and based in rue Monge, the core of Paris's Fifth Arrondissement, a popular neighbourhood where Grave also lived and established the redaction of the journals he edited. In Ellis's vivid account:

> Still less scholarly, but more original, and really an artist in style, was Élie Reclus, the brother of the better-known Élisée and the uncle of his namesake Élie Faure, the historian of art and likewise an artist in style, who was later also to be my friend. Élie Reclus was a delightful personality, simple and genuine, an Anarchist of the finest quality. I have always cherished a warm regard for him. He invited me to dinner to meet his family and some special friends. I fear they were much disappointed, for not only did I keep dinner waiting by losing my way, but I do not shine on such occasions even when free to speak my own language.[188]

The transnational nature of Élie Reclus's figure did not escape Ellis's observation: it was in his house that the Kropotkins had found refuge in January 1886 after Pyotr's release from prison and before their departure for England. It was there that Ellis felt free to have conversations in English, accordingly a rare facility in Paris at that time, and that he met a certain number of anarchist refugees.

> Élie Reclus's house was a place of refuge for distinguished foreign Anarchists in hiding. I remember meeting one there once, [Francesco Saverio] Merlino, a gloomy, silent, dark little man. Later, Élie Reclus found it convenient to migrate to Brussels. When in after years my studies were proscribed and condemned the finest and most affectionate letter that reached me was from Élie Reclus.[189]

In these few lines by Ellis it is possible to find a summary of the transnational anarchist networks, their mobility, multilingualism and concern with solidarity and free speech.

Ellis was also the author of a tribute to Élie Reclus in Ishill's collective book, providing a rather detailed biography of the French scholar. It is worth noting that Élie was much less famous than Élisée and that even in French literature studies of Élie's life and works are rare. Beyond their personal acquaintance, Ellis's interest in this figure was an intellectual affinity and some sort of fascination for this example of 'fighting intellectual'. Ellis gave special emphasis to Élie's experience during the 1871 Paris Commune, when he was appointed as the director of the National Library of France. Against the mainstream press claiming that the 'barbarous' of the Commune were about to destroy the cultural patrimony of the *Ville Lumière*, Ellis stated how: 'During those troublous days, when these most precious collections were exposed to dangers of all kinds – bombardment, incendiarism, the ignorant hatred of social fanatics – Élie Reclus was indefatigable; he adopted every available precaution, took measures to save the *Bibliothèque* from the great fire'.[190] Ellis was especially interested in Reclus's anthropological studies, comparing his work to contemporary British scholarship and especially with 'Frazer, Hartland, and Crawley. [Élie Reclus's] position was well recognized in England. A number of the articles in this department of the *Encyclopaedia Britannica* were from his hand, and his best book, *Les Primitifs*, translated into English by Mrs. Wilson under the title of *Primitive Folk*, was included in the *Contemporary Science* series'.[191] This series was edited by Ellis: like contemporary works on sex by Ellis, Thomson and Geddes, Élie's book was regarded as 'shocking by the late Victorian public'.[192] For Ellis, 'the special characteristic of his scientific work is its sympathy with the operations of the savage mind and its insight into the working of savage intelligence'.[193] Though Ellis considered Élie Reclus's findings as outdated in comparison to what he defined as the 'rigour' of his days' science, he was interested in Reclus's pioneering efforts to see first peoples not as inferior or less intelligent than the so-called civilised ones, but to consider their societies as different strategies to deal with different environments, which did not imply a mechanistic geographical determinism, as the emphasis was always given to human agency, analysing especially social and political practices of cooperation. 'Élie Reclus was one of the first to show that savage beliefs and customs, however outrageous or fantastic they may appear from the civilized point of view, have a demonstrable reasonableness and a justifiable morality when studied in connection with their environment.'[194]

Beyond his admiration for Élie Reclus's person, Ellis also recalled this figure as an exponent of transnational solidarity networks and of what might be deemed as a 'humanist', politically persecuted for his ideas.

> Those who knew him best loved him most; 'he was a man of infinite sweetness and goodness', truly writes to me one of his colleagues at the *Université Nouvelle*. Although he lost his early faith in revolutionary action, he never lost his sympathy with those who suffered, rightly or wrongly, in the cause of humanity; in his little apartment in Paris, and afterwards in Brussels, men of science mingled with the outcast practical idealist of all nations. In the morning you might read in the papers how some revolutionary leader was being hotly persecuted by a foreign Government; in the evening you might find him in the person of some pale, silent little man sitting by the fire in Reclus's salon. For every brave word or deed, in any country, which seemed to him a blow struck for social or intellectual freedom, Élie Reclus was up to the last full of a tender, appreciative, unworldly sympathy which could never be forgotten by those who had experienced it. ... He was one of those men who are ranked among the criminals while they live, among the saints when they are dead.[195]

Likewise, Ellis published a tribute to Kropotkin in the book edited by Ishill in 1923, where he greatly praised the person, albeit criticising him for some of his old-fashioned neglecting of the rising psychological sciences. These writings, and three letters sent by Ellis to Kropotkin to request his editorial collaboration for the series 'The Life of Nations', surviving in Moscow, can account for their acquaintance.[196] Ellis's 'great work on sex' was endorsed by Emma Goldman in her autobiography[197] showing that Ellis's scholarly and civil work was not far from the preoccupations of anarchists at that time, at least the more intellectual and sensitive to the topics of sexual liberation and 'homogenic love', which were widely addressed from the 1950s by famous French anarchist Daniel Guérin (1904–1988) in the context of the sexual revolution.[198]

Ethical socialism: Henry Salt, William Morris and Walter Crane

In the complex activist British milieus at the time of Kropotkin's return to England after his jail term in France, the relation between socialism and 'human's inner life' was supported by groups such as the Fellowships of the New Life, in which Ellis, Carpenter and Salt participated in varying degrees. When Morris and the Socialist League broke with Hyndman and the Social-Democratic Federation, 'Carpenter and his circle inclined towards them'.[199] Eventually, Salt[200] and Carpenter[201] were contributors to *Commonweal*, and they both subscribed to the Socialist League for some time.[202] I would argue that the heterogeneous range of authors gravitating to Salt's, Carpenter's and Morris's social circles had important relations with anarchism, especially from an intellectual prospective. These relations are revealed by their correspondences with Kropotkin, Reclus and their involvement in *Freedom*. I also contend that these collaborations can be considered a key component of a broad definition of 'ethical socialism', making sense of the intersections which existed between anarchist geographers' work and traditions of British socialism. Some of the most significant aspects of these links are exposed by the exceptional experience of Salt's *Humane Review*, published from 1900 to 1910.

Henry Salt (1851–1939) was a close friend of Carpenter and an early advocate of the abolition of the death penalty,[203] of vegetarianism and of animal rights.[204] His range of occupations also included anti-militarism, significantly exemplified by his criticism of the 'Fabian support of the Boer war'.[205] Recently, Anna Clark has stressed the originality of Salt's and Carpenter's Humanitarian League in the wider panorama of 'mainstream' humanitarianism because, for instance: 'Unlike the Fabians and New Liberals, the Humanitarian League opposed imperialism because it violated human rights'.[206] The breadth of Salt's interests, endeavours and networks reveals a complexity of thinking which still deserves further investigation in his attempts to question anthropocentrism, 'regaining the unity of man and nature',[207] through which Salt 'transcended the dichotomy of society and nature'.[208] George Hendrick observed that Salt was familiar with different socialist circles in London thanks to his brother-in-law Jim Joynes, who 'was able to work with many opposing organizations; for example, he knew and worked with both Hyndman and William Morris. Salt was welcomed into these circles, and he met such reformers and revolutionaries as Sidney Olivier, William Morris, Graham Wallas, the Webbs, Hyndman, Bax, and Prince Kropotkin'.[209]

In his autobiography, Salt referred to his acquaintance with both Kropotkin and the Recluses. When he lived in Surrey, he met Kropotkin who was then looking for a house, which he would eventually find at Viola Cottage, Bromley, Kent. On that occasion, the two men 'spent a day in walking about on that quest'; after that experience, which took place accordingly between 1893 and 1894, Salt could provide an exemplar anecdote on Kropotkin's habits and on the reasons for their familiarity.

> We met a troop of beggars whose appearance was decidedly professional; and I noticed that Kropotkin at once responded to their appeal. Later in the day we fell in with the same party, and again, when they told their tale of woe, Kropotkin put his hand in his pocket. At this I ventured to ask him whether he had observed that they were the same lot; to which he replied: 'Oh, yes. I know they are probably impostors and will drink the money at the public house; but we are going back to our comfortable tea, and I cannot run the risk of refusing help where it may possibly be needed.' If in this matter one sympathizes with Kropotkin rather than with the Charity Organization folk, I suppose it is on Shelley's principle—that he would 'rather be damned with Plato and Lord Bacon than be saved with Paley and Malthus'.[210]

Significantly, Salt understood Kropotkin's behaviour in the terms of his own British cultural position; most importantly, he appreciated this example of applying one's ethical principles in daily life, which accordingly matched his own views.

Salt considered Reclus's work as inspirational. First, in Salt's autobiography, the chapter on humanitarianism was opened with a citation from Reclus arguing that 'The barbarian gives to the earth he lives on an aspect of rough brutality'.[211] This book was titled *Seventy Years Among Savages*: this title recalled very directly the ironic overturning of the term's meaning mobilised in the works of early anarchist geographers in criticising colonial crimes, by attributing the qualification of true 'savages' and 'barbarous' to the conquerors, as discussed in Chapter 4. Ironically, in this case Salt used the definition of 'savages' for defining British society at this time, with special reference to the habit of meat-eating, which Salt often compared with cannibalism. A vegetarian and an early 'environmentalist', Reclus was mentioned by the British author as an example of applying 'humanitarian' principles to the generalised commodification of mountain milieus.

> Humanitarianism is not merely an expression of sympathy with pain: it is a protest against all tyranny and desecration, whether such wrong be

done by the infliction of suffering on sentient beings, or by the Vandalism which can ruthlessly destroy the natural grace of the earth. It is in man's dealings with the mountains, where, owing to the untameable wildness of the scenery, any injury is certain to be irreparable, that the marks of the modern Vandal are most clearly seen. It so happens that as I have known the mountains of Carnarvonshire and Cumberland rather intimately for many years, the process of spoliation which, as Élisée Reclus has remarked, is a characteristic of barbarism, has been there forced on my attention.[212]

In the *Humane Review*, in 1909, Salt argued against the commodification of the mountain landscape for building tourism infrastructures, which he identified with the results of landlordism and of the 'vandalism of trade',[213] corresponding with an argument that Reclus developed in his earliest geographical writings.[214]

Reclus, Kropotkin and Carpenter were not the exclusive anarchist inspirations of Salt, who included in this category (beyond Tolstoy, whom he admired) another collaborator of the *Humane Review*, 'Ernest Crosby [1856–1907], another philosophic anarchist, [who] was perhaps as little known, in proportion to his great merits, as any writer of our time'.[215] My survey of the *Freedom* collection shows that Salt was both directly and indirectly involved in the journal. In 1895, he published in *Freedom* the important paper 'Respectability and its martyrs', a reprimand against the social conformism of Victorian society, discussing how the bar of respectability was set up between social prestige and social blame. Salt defined respectability as a 'perverted term', arguing that its practice, in terms of dressing, exterior look and social habits, implied a daily mortification of the flesh worthy of the martyrs of the religions, or of the so-despised 'savages'. 'The endurance shown by the Respectables, in their compliance with the ritual of their creed, may be aptly compared with the behaviour of the so-called "savages" of the Pacific, who patiently submit themselves to the analogous process of tattooing—though it may be questioned whether it is not less savage on the whole, to be tattooed in body than in mind'.[216] Yet, the main danger represented by these social conformists was not for themselves, but for the others, constantly judged and compelled to obey respectability laws so as not to be excluded by society.

Inhumanity to others is implied, not directly or consciously perhaps, but nonetheless purely, in their creed. For Respectability is the foe of all naturalness and instinct; it is the determination to keep up appearances at all costs, without the slightest care for the actualities of compassion, humaneness, and justice; so that Shelley was right when he described 'the respectable man' as 'the smooth, sniffing, polished villain; whom all the city honours; whose very trade is lies and murder; who buys his daily bread with the blood and tears of men.' Heartlessness towards others, joylessness in himself—those are the two dominant qualities of the nineteenth-century Respectable.[217]

These lines might contain implicit references to sexual regulations as part of this mortification, anticipating the cases raised by Carpenter, Ellis and others in the same decade. In any case, they represent a landmark for the interest of anarchists and early socialists in human inner life, daily behaviours and feelings rather than dealing only with economic issues.

Salt greatly admired Percy Bysshe Shelley, and he published a biographical study of the poet, which was reviewed by Marsh, in *Freedom*, in July 1896, under the title 'The poet of freedom'. Being the son-in-law of William Godwin and Mary Wollstonecraft, Shelley had all the requirements to be popular among anarchists who, according to Marsh, 'should study the life and writings of this great poet', although English literature did not lack writers such as 'Byron, Wordsworth [who] had their dreams of liberty'.[218] According to Marsh, Salt's merit was to highlight social critiques in Shelley's work, explaining the blame that the poet received by some of his contemporaries who considered his production as 'demoniac', to the point where Marsh included Shelley among the forerunners of anarchism. Marsh also appreciated how Salt addressed Shelley's ideas on sex: 'He presents them to the public freed from the lies and slanders with which a base conventionality and a sickening hypocrisy has striven to entangle them. All Anarchists and, we should like to add, all sensible and unprejudiced persons who have given thought to the subject will thank Mr. Salt for this'.[219] Thus, the editor of *Freedom* took public position in favour of disclosing 'scabrous' matters beyond the social hypocrisy willing to censor the 'libertine' attitudes of authors like Shelley, in the aftermath of the Wilde case. Only on one point did Marsh take some distance from Salt's positions, arguing that blame for the 'Harriet affair' deserved not to be placed on Shelley, but on society. Harriet Westbrook (1796–1816), first wife of Shelley, committed suicide after her husband's escape with Mary Godwin but, according to Marsh, what deserved greater consideration in that account was how 'society could be so criminal as to force poor Harriet into the course which was her ruin'.[220] This was an implicit reference to the problem of divorce, for which the Recluses had indirectly campaigned in France through their support for free unions,[221] confirming the *Freedom* group's commitment to civil rights.

In February 1897, *Freedom* organised a mass meeting at the Club and Institute Union Hall against the torture of Spanish anarchists in the fortress of Montjuich, near Barcelona. Among the speakers, Kropotkin evoked the infamous memory of the Inquisition, and messages of support came from Salt, Carpenter and Crane. Salt pointed out the 'rumour of Spanish outrages' in still colonised Cuba, but most importantly, the leader of the Humanitarian League broke the refrain of British scholars and socialist politicians who expressed solidarity with foreign anarchists claiming that Britain was always a freer and more friendly country to them. Evoking the issue of death penalty, Salt argued that: 'A nation which treats its own prisoners as barbarously as we English do, has small right to remonstrate on other countries on the use of torture. But if Cruelty is international, this is all the mere reason why

Humanity should be international also, and why a protest of this sort should be made as worldwide as possible'.[222] The definition of humanity was likewise used by Carpenter and Crane in their respective notes, showing a rather smooth connection about 'universal' humanistic concerns and anarchist internationalism. In the following years, *Freedom* advertised initiatives of the Humanitarian League like the lectures of pacifist and atheist militant Hypathia Bradlaugh Banner (1858–1935) against corporal punishments, practices defined as 'barbarous'[223] in the anarchist journal. In July 1913, an anonymous article on 'Clothes and Morality'[224] returned to some elements of Salt's earlier critique of respectability, referring to the possibility of naturism, an option famously endorsed by the late Élisée Reclus.[225]

This wide range of topics was addressed by *The Humane Review*, a journal which Salt launched in 1900 as a way for deepening theoretically the discussions carried on in the other journal of the Humanitarian League, *Humanity*, published from 1895 to 1914. *The Humane Review* started its publications with a statement rather close to the concerns of many anarchists: 'The League is doing work which, though it must commit itself to all humane persons, is not undertaken by any other society; for no other organised protest has ever been raised against Cruelty in all its forms'.[226] Therefore, humanitarianism was distinguished from philanthropy because it included non-human beings in its concerns. If *Humanity* looked more like a bulletin, the *Humane Review* had space for longer, and more reflective papers.

In the opening issue, Salt clarified that this journal had been founded

> to give expression to an aspect of life which has not yet received, and is not yet likely to receive, its due share of attention. The barbarity of which, in nominally civilised societies, human beings are too often the victims, together with the widespread disregard of any moral obligation to the lower animals, shows what great need there is of a humane, as well as a merely academic school of ethics; and it is hoped that some service may be done to the cause of humanity.[227]

Then, Salt outlined where these problems were located, expressing an early idea of the interconnection of different forms of oppression, and therefore of different forms of resistance, close to what is today defined as intersectionality.

> First, such national and social questions as peace and arbitration, the treatment of native races, the sweating system, the criminal law and prison system, capital and corporal punishments, the municipalisation of hospitals etc.; and secondly, the various problems relating to the treatment of animals, as in vivisection, blood-sports, the fur and feather trade, and the cattle-traffic, which subjects will be regarded as part and parcel of the social question, and not as a separate and subordinate branch of it.[228]

One might add to this list women's emancipation, considering the papers published by Margaret S. Clayton, member of the Women's Freedom League together with her husband Joseph Clayton, likewise a contributor to the *Humane Review*, and the Nevinsons. The unpublished letters which Margaret sent to Carpenter at the occasion of the celebrations for Kropotkin's 70th birthday show that the Claytons were Kropotkin's friends and that they collaborated actively with Carpenter in involving people to sign the address reproduced in Appendix C.[229] Margaret wrote a first paper in the *Humane Review*, relating her own experience of police repression when she was arrested after a suffragist protest.[230] Her second paper was dedicated to the figure of Mary Wollstonecraft, whom Clayton praised for her way of associating feminism, antimilitarism and opposition to the death penalty. For Clayton, Godwin's wife inspired likewise Josephine Butler.[231] Again, humanitarians paralleled subject matters raised by the anarchist geographers.

Since its first year, one of the characteristics of *The Humane Review* was an explicit anti-racist and anti-colonial critique of European civilisation, represented by a series of papers including Heath's text 'The Kafir and his master' discussed on p. 166. The Secretary of the Aborigines Protection Society Henry Fox Bourne (1837–1909) wrote on 'The claims of uncivilised races' countering the pretentions of 'nations calling themselves civilised'[232] on people which he simply considered as 'different', condemning the 'deplorable' acts which led to 'the ruin of aborigines of Australia, New Zealand and nearly all the smaller islands in the Southern sea'.[233] If these protection societies were often considered as too paternalistic by anarchists like Reclus,[234] Bourne's declarations were in general radical enough to be compared with the other anti-colonial materials analysed in this book, for his claims that colonised peoples should have all rights on *their* land. An article of Carpenter analysed the situation of the British Empire in India, where the author had had the occasion to travel. Carpenter was accordingly one of the rare Victorian authors denouncing famine and starvation in the colonies, a set of phenomena which are called today 'Victorian Holocausts' by Mike Davis.[235] But what was significant and anticipated more recent scholarship was the public acknowledgement of the colonial nature of that drama, when Carpenter stated that Indian people were 'dying by millions and millions *of* our sovereign and imperial neglect'.[236] Carpenter blamed the British occupation of India under different standpoints: first, in its religious and cultural aspects, by denouncing the complicity of missionaries in the conquest of the region;[237] second, he fumed at the predatory aspects of British rule on Indian people, by arguing that 'we have our hand perpetually in their pocket';[238] third, he matched a classical argument by Reclus and Kropotkin against colonialism in different countries (including Ireland) and capitalist industrialisation in general by observing that, in India, 'we have practically destroyed the village community'.[239] Carpenter related his personal experiences of talking with Indians, who regularly recounted the arrogance of his compatriots in their daily relations with the 'natives', and concluded with remarks which applied not only to the British one, but to all colonialism. 'The cry of Empire is the crazy cry of imbecile and tottering authority, not only in England but in all the government-bestridden nations of the West ... These fatuous empires... will fall and be resent asunder. The hollow masks of them will perish. And the sooner the better.'[240] The tone of the invective and Carpenter's prediction of decolonisation clearly recalled Reclus's statements quoted in Chapter 4.

Translated by Carpenter, Reclus was one of the key authors of the *Humane Review*. His paper 'On vegetarianism' was published in the first issue and inaugurated a long series of contributions by several authors against meat eating, blood sports, feather industry and other forms of animals' killing and exploitation. Reclus's text has been rediscovered and valued in the last years in connection with the growing reflections on vegetarianism and veganism in the field of anarchist geographies, including the creation of a 'Vegan geographies collective', in the context of a wider discourse on animal rights.[241] In particular, Richard White defines this text by Reclus as an inspiration for 'an expanded anarchist geographical praxis that embraces more than human animals when speaking of ethics and social justice',[242] including aspects of everyday violence at the micro scale, connecting abuses on animals and abuses on persons. Reclus's second article was 'The Great Kinship', explaining his ideas of human–animal solidarity and association by considering early stages of human history not as less progressed, but as more sensitive to the non-human world, because 'primitive man was thinking of a fraternal association [with animals],'[243] rather than their domestication. Here, anarchist anti-authoritarian principles were potentially applied to all sentient beings. If respect for animals might repose on a mainly ethical stance, in Reclus's case it was overtly politicised, starting from his terminology. For instance, he defined the Amazonian big houses sheltering both humans and animals as 'this whole republic',[244] considering that the presence of animals (and not only pets) in a house would contribute to a 'broad atmosphere of peace and love'.[245] Reclus's paper ended with a tale about a bull and a zebu: the first animal was angry and dangerous because he was treated in a European way, that is bounded and beaten by his master; the second was peaceful and collaborative because the person who worked with him, an Indian, treated his animal in a loving way.

At Reclus's death, the *Humane Review* published Heath's obituary of him mentioned on p. 168, and the Humanitarian League republished Reclus's essay as an independent brochure, together with another text on vegetarianism, 'The Meat Fetish' by Ernest Crosby.[246] This metaphor of the fetish was connected to one of the favourite Salt's polemic arguments, namely the comparison between the European habit of consuming meat and the true or alleged cannibalism which the tenants of the 'superior civilisation' attributed to 'savage' tribes. In a paper titled 'concerning cannibalism', Salt displayed his humorous side by presenting under the same light what he called the 'English Roast Beef' and the 'African Roast Man'.[247]

Among the other authors of the *Review*, Anne Cobden-Sanderson published a paper on 'Domestic Economy', in which she criticised the gendered division of labour under the standpoint of well-being and personal relations in the household.[248] Among Reclus's and Kropotkin's acquaintances, both Richard Heath and his son Carl contributed several papers for Salt's journal. One of the most significant was a text by Carl Heath protesting upon the execution of anarchist educator Francisco Ferrer by the Spanish state. Francisco Ferrer y Guardia (1859–1909), both Reclus's and Kropotkin's friend, is considered as one of the big international inspirers of the movement of anarchist education.[249] His trial, an overtly political one, was the occasion for an international gathering of anarchists and other progressives to request (unsuccessfully) his liberation. Heath evoked first this international protest against 'the soul-destroying tyranny of priest and soldier',[250] matching the anticlerical tones of the anarchist propaganda, which focused on the strong interference the Catholic Church exerted in that affair against 'all liberty of thought'[251] to counter secular education. Heath made this criticism from a different perspective than the atheist position of militant anarchists, but one might consider significant his references to the French secular school system and the radical French politician Alfred Naquet (1834–1916), an acquaintance of Reclus. Naquet would be the proponent of the French law on divorce as an MP, at a time when French anarchists collaborated with the project of secular schooling by radical protestant Ferdinand Buisson (1841–1932) for the common task of getting rid of any religious interference in the public school.[252] An unorthodox protestant like his father, Heath concurred with internationalist and anti-militarist remarks, defining the concept of military honour as 'synonymous with every form of tyranny, of lawlessness, of blood [because] democracy acknowledges no foreign nations. Boundaries and armaments are the weapons of finance and plutocracy all the world round. Ferrer was our friend because he and we had become conscious of that new spirit of human solidarity, of universal kinship'.[253] This is my last example of how the *Humane Review* could favour the gathering of a heterogeneous array of scholars and activists sharing radical criticisms of society which started from for daily life. Anarchists and anarchist topics found their place in this intellectual environment.

Many of Kropotkin's concepts, such as mutual aid, were also set out in the Humanitarian League's publications. His personal acquaintance with Salt, mentioned in Salt's autobiography, is confirmed by seven of Salt's letters surviving in Moscow, where Salt called Kropotkin 'dear comrade'.[254] The breadth of Salt's interests in the anarchist field is also well expressed by his recollections on Élisée Reclus, in which he stated that the French anarchist geographer helped the League financially.

Count Tolstoy, it goes without saying, was in full sympathy with us; and so was that many-sided man of genius, M. Élisée Reclus. Famed as geographer, philosopher, and revolutionist, one is tempted to sum him up in the word 'Poet'; for though he did not write in verse, he was a great master of language, unsurpassed in lucidity of thought and serene beauty of style. He was a vegetarian, and the grounds of his faith are set forth in a luminous essay on that subject which he wrote for the Humanitarian League. Very beautiful, too, is his article on 'The Great Kinship', worthily translated by Edward Carpenter ... His anarchist views prevented him from formally joining an association which aimed at legislative action; but his help was always freely given. 'I send you my small subscription', he wrote, 'without any engagement for the future, not knowing beforehand if next year I will be penniless or not.' I only once saw Élisée Reclus; it was on the occasion of an anarchist meeting in which he took part, and he then impressed me as being the Grand Old Man.[255]

It is worth noting that, in general, anarchists are not aprioristically opposed to what is related to 'legislative action'; only, they try to distinguish between social 'good causes' and their possible exploitation by politicians.

In one of the last published issues of the *Humane Review*, in April 1910, Salt had the occasion to pay a tribute to William Morris (1834–1896) in a paper on socialism and poetry, arguing that Morris's 'wonderful stanzas ... might redeem socialism from the charge of being unpoetical'.[256] Morris was one of the first and more famous acquaintances of Kropotkin in Britain: if his association with Kropotkin was already discussed in the literature mentioned above, especially by Fiona McCarthy and Ruth Kinna, I would extend these invaluable works by reconsidering some aspects of this association, drawing upon newly available sources such as Morris' letters surviving in Moscow and my surveys of *Freedom* and the *Humane Review*. I especially contend that Morris's association with anarchism can be read in the context of the humanitarian traditions of British socialism and its parallel involvement with culture and education, which clearly interested figures like Reclus, Kropotkin and their entourage. In 1885, Morris welcomed Carpenter into the Socialist League, while receiving from him an invitation to lecture in Sheffield.[257] After his visit to the Millthorpe community, where Carpenter lived, Morris observed that 'Carpenter seems to live in great amity with the workmen and the women, they all live together in the kitchen, and this is very pleasant'.[258] The founder of the Arts and Crafts movement likewise corresponded with Salt.[259] Without doubt, Morris can be considered as an advocate of 'humane sciences' including struggles for civil and social rights beyond the traditional spheres of socialist politics.

Morris and anarchism

Some early biographies of Morris, like the one by Henderson, pay little attention to Morris's relations with Kropotkin, though highlighting his friendship with Carpenter and also relating some spicy anecdotes on the nickname Carpenter was given in the London circuits of the Socialist League

and the Fabians: 'The Noble Savage'.[260] On the Marxism side, Morris's biography by E.P. Thompson was biased by strong anti-anarchist prejudices.
Arguing teleologically for Morris's progressive evolution from Romanticism
to the 'true' socialism (accordingly Marxism), Thompson struggled to understand why, in the mid-1880s, Morris got closer to anarchism despite that 'it
would be wrong to think of the "Anarchists" within the League in 1885 and
1886 as the conscious advocates of certain theoretical positions'.[261] Apparently, everything which did not engage with theoretical foundations was not
considered as a valid position by Thompson, who suggested that, towards the
end of that decade, Kropotkin's growing influence was due to his 'romantic
history even more than his writings'.[262] Such was Thompson's commitment to
understanding anarchism that he classified British left-libertarians of that
time, excluding the so-called 'moderate' such as Kropotkin and Carpenter, as
'a curious assortment of cranks and fire-eaters'[263] without considering that,
on the standpoint of more orthodox Marxists than Thompson, even Morris's
contributions could be discarded on similar grounds. However, Thompson
appeared a rather lone voice in stressing the Marxist influences on Morris to
undermine the anarchist ones. Authors like G.D.H. Cole, James Hulse, Peter
Marshall[264] and more recently David Goodway, have highlighted the anarchist side of Morris's thought.[265] Kinna opportunely investigates the importance of Romanticism and art in informing Morris's special visions of
socialism,[266] while MacCarthy defines Morris as 'an educational anarchist in
the tradition of Godwin, Kropotkin and Goodman'.[267] I would argue that no
definition, including the 'anarchist' one (moreover explicitly rejected by
Morris), could capture the complexity of Morris's figure, but anarchism is
undoubtedly a relevant lens through which to read Morris's work, and his
association with anarchism still needs further investigation. To fill this lacuna,
Morris's engagement with anarchists, in the context of the fluidity and porosity
of early British socialist milieus, will be discussed here.

Morris was a protagonist, together with Charlotte Wilson, of the split
which occurred in the Fabian Society in 1886 about parliamentarianism. On
the one hand, many members joined the positions of Hyndman's Social
Democratic Federation favourable to electoral struggle; on the other, anarchists
and members of Morris's Socialist League argued that

> Whereas the first duty of Socialists is to educate people to understand what
> their present position is and what the future might be, and to keep the
> principles of socialism steadily before them; and whereas no Parliamentary
> party can exist without compromise and concession, which would hinder
> that education and obscure those principles: it would be a false step for
> Socialists to attempt to take part in the Parliamentary contest.[268]

In her recollections, Wilson accounted for Morris's material support to the
Freedom publishing endeavours: 'From its first appearance [until 1888], *Freedom* was set up at the Socialist League printing office by the kind permission

of William Morris'.[269] It is also worth noting that the journal launched by
Morris in 1885, the *Commonweal*, counted on the participation of anarchists
like Nettlau and always maintained a certain proximity with *Freedom*, especially after the *Commonweal* was resumed by an editorial group strongly
inspired by anarchism, as discussed by John Quail.[270] Finally, the *Commonweal*
and *Freedom* groups merged in 'April/May 1895'.[271]

Nevertheless, Morris always refused the definition of 'anarchist', taking a
clear and explicit political distance from anarchism, though in the terms of
mutual respect and solidarity. This reinforces the idea that the heterogeneous
milieu of diverse scholars and activists frequented by Reclus and Kropotkin
in Britain could collaborate on the bases of shared ethical concerns maintaining the respective ideological differences. Morris collaborated indirectly
with *Freedom*, for instance by delivering a benefit lecture for the *Freedom*
publication fund in March 1893, in coordination with a speech by Kropotkin:
in the first week, the Russian geographer lectured on 'Anarchism', and in the
second one Morris presented on 'Communism'.[272] Morris's lecture, published
in *Freedom*, clarified that, rejecting the imposition of a majority, anarchism
did not fit his idea of social change, as he argued that this change should be
imposed to someone at a certain moment. Anyway, he confirmed his empathy
with the *Freedom* group and their egalitarian aims.[273]

At Morris's death in 1896, the first page of *Freedom* was dedicated to him,
with tributes by Crane, Kropotkin and John Kenworthy, plus a collective
message from some of 'his old comrades'. In particular, the Anarchist Prince
insisted on the uniqueness of Morris's figure in the socialist field, acknowledging his work as an inspiration for considering human feelings as part of
the revolution: 'He was almost the only poet of the joys of life, the joys which
man finds in the conquest of freedom, in the full exercise of all his powers, in
work – the work, the work of his hands and his brain … the happiness that
men can find in conquering full freedom and freely associating with their
equals'.[274] Therefore, it can be argued that Morris played a role in keeping a
significant part of British socialism away from economic based theories, like
classical Marxism, and closer to daily life, which implied points of contact
with both anarchism and humanitarian socialism. At the end of his paper,
Kropotkin highlighted his friend's proximity with his own views.

> For the last few years of his life, Morris had abandoned the Socialist
> movement, and he frankly explained his reasons in a lecture which he
> delivered for the anarchists at Grafton Hall in 1893. If the movement had
> gone on developing and bringing England to a Social Revolution, Morris
> undoubtedly would have gone under the red flag as far as the masses
> would have carried it. But the endurability of the workers, who patiently
> support any amount of capitalist oppression, deeply affected him. More-
> over Morris, who would have gone any way with the masses, could not go
> with parties.[275]

This indicated some of Morris's political and intellectual proximity with the *Freedom* group, an impression reinforced by a Carpenter's tribute published in the following issue, which focused mainly on Morris's character, thus revealing a profound friendship between the two men.[276]

More reminiscences on Morris were published in the following issues of *Freedom*. [277] In 1911, an indignant note responded to an article by socialist teacher Harry Lowerison (1863–1935), who reported an episode in which Morris allegedly ridiculed a group of anarchists who were whistling him. The arguments of the *Freedom*'s editors clarified their view of the relations between Morris and the anarchists.

> Morris had his quarrel with Anarchists, as he had with plenty of other people, and as he would have with Lowerinson could he read the fulsome nonsense quoted above. The statement is wholly misleading, as it gives the impression of a bitter antagonism to Anarchism which Morris never had. He did not like the name, and he never would admit the logical deductions as to government and authority upon which Anarchists insisted. But his ideas came very near to ours, and when he published *News from Nowhere* everyone said it was a picture of Anarchist Communist society in all but name. Besides, we know personally how courteous and fair he was in letting the Socialist League Hall for our first meetings. One of his last public lectures was given for the benefit of *Freedom* with the utmost goodwill.[278]

The point about *News from Nowhere* was part of a long tradition of positive anarchist reception of Morris's book. One of Marsh's successors as the editor of *Freedom* in the following decades, Marie-Louise Berneri (1918–1949), wrote a history of utopias in which she deemed *News from Nowhere* as one of the rare utopian worlds which can be deemed genuinely libertarian, as importance is given to individual liberties.[279] On the contrary, works like *Looking Backward* by Edward Bellamy were considered as authoritarian utopias where the individual was only a pawn of an all-powerful state, as already observed by Kropotkin in his obituary of Bellamy.[280]

As for Kropotkin's experience, Morris first met him 'in a meeting in commemoration of the Commune at South Place'[281] in March 1886, and very much liked the Russian exile. In the second half of the 1880s, Morris and Kropotkin corresponded for the organisation of Kropotkin's lectures in Hammersmith and other places where Morris was able to organise them; in 1888 one was held in the West Kensington Club, in a room able to contain 'over 300'[282] attendants. Surviving in Moscow and only partially reproduced in the invaluable *Correspondence* edited by Norman Kelvin, Morris's letters to Kropotkin reveal that their collaboration included commitment to political prisoners. In a letter from 1892 (included in the published edition), Morris joined the campaign for supporting the appeal of French anarchist exile Jean-Pierre François (1855–), convicted under dynamite allegations, whom the

British authorities tried to extradite in France, triggering a heated debate on the defence of the right of asylum.[283] Morris joined the campaign, for which Kropotkin also requested his help to find a suitable journal for publishing related materials.

> I fear that I have no influence with the papers except it might be the *Chronicle*. Of course, I quite agree with you as to the desirability of getting the appeal tried: as it is obvious that if the French police can get a man extradited by merely trumping up any pretence at evidence against him, no one is safe. I have already subscribed ... but I will send Miss [Agnes] Henry another subscription of £5 today. I have much hope of the judges reversing the magistrates' decision.[284]

This was what happened, consistent with other discussions between Morris and Kropotkin on the question of prisons, another point they shared with the humanitarians. As Morris wrote to Kropotkin: 'There are perhaps some points of disagreement between us, but on this matter of prisons I most heartily agree with you'.[285]

An extreme frankness characterised the exchanges between the two men, who did not hesitate to address their respective disagreements, but who were also clear in stating their points in common. An unpublished letter where Morris invited Kropotkin's contribution for *Commonweal* contained perhaps his most explicit declaration of some proximity with anarchism. 'If you would from time to time contribute an article to the *Commonweal*, I think we may at least claim for it that it is not exclusive, and that it recognizes the necessity for keeping up the revolutionary spirit.... We agree with you in the main, though the League has not formally called itself Communist Anarchist.'[286] Morris likewise supported *Freedom*'s material distribution: for instance, in 1891, he ordered a certain number of copies to Nannie Dryhurst,[287] with whom he also corresponded on professional matters.[288]

Kropotkin's lectures in Hammersmith, which were regularly the occasion for invitations to dine and sleep in Morris's house, addressed topics like 'Socialism and its modern tendencies'.[289] The confidence between the two men is demonstrated by the abundant information Kropotkin gave on domestic matters, especially after the birth of his daughter Alexandra (Sasha) in 1887. Kropotkin also corresponded with Morris's daughter May on editorial issues and on the organisation of these lectures. An anecdote reveals some material aspects of Kropotkin's approach to his own solidarity networks. In his first years in Britain and before the definitive commercial success of his books and articles, the Anarchist Prince had pressing material needs, and lectures were also a way to integrate his income, because the public paid to attend them. In Hammersmith, the right to attend was established as '£1 the ticket'.[290] Woodcock and Avakumović recount that a group of Kropotkin's friends gravitating around Morris organised a series of six well-paid lectures at the Kensington Town Hall with the main objective of helping the

Russian exile. Unluckily, this purpose was displayed too explicitly by some-one, so that the Anarchist Prince 'smelt charity, felt his dignity attacked'[291] and cancelled the lectures. A letter sent to May Morris contained the details of Kropotkin's arguments. After long and even tedious declarations that this decision was made after long consideration and was absolutely definitive, the Russian geographer explained:

> After having taken knowledge of letters written in connection with them, my personal dignity, as well as the interest of the cause, forcedly com-mand totally to abandon the idea of the lectures. I am extremely sorry to come to that decision because it will be disagreeable to you and the friends who joined to help you. I am extremely sorry to make to you this disagreement, because I know how you took it to heart. But I am per-suaded that if you knew the correspondence written in connection with that, you would be the first to [agree].[292]

Confirming his 'best sympathies' to Miss Morris, Kropotkin even offered to refund the organisation from his own pocket in case of technical incon-veniences. This anecdote has a twofold meaning: first, it shows how proble-matic were questions of wage-earning and concrete solidarity among exiles and activists; second, it shows the importance of sociability networks to understand choices and orientations of their actors, especially considering that in those networks not only material business, but also symbolic and moral capital mattered. It would be easy to read in Kropotkin's behaviour a result of his aristocratic education (which might partially explain it), but his reference to the 'cause' suggests that Kropotkin feared that being identified as a money seeker would have undermined his political credibility. There-fore, he considered it better to continue earning his life in the editorial market, exploiting bourgeois assets without soliciting the socialist movement for that.

Finally, Kropotkin and the *Freedom* group had important relations with one of Morris's close collaborators, Walter Crane (1845–1915), one of the protagonists of the Arts and Crafts movement. Though Crane is not remem-bered as an anarchist or as a geographer, his trajectory intersected sig-nificantly with both geography and anarchism. In a 2010 article, Felix Driver addresses the figure of Crane as the author of the famous Imperial Federation Map of the World,[293] showing the extent of the British Empire in 1886 and often used as the symbol of imperial mapping.[294] Amazingly, the authorship of this map was overlooked by all critics until a paper by Pippa Biltcliffe identifying Crane as its author was published in *Imago Mundi*.[295] According to Driver, this long under-consideration of the relations between socialism and Empire was related to scholars' difficulties in appreciating visual materi-als, which might upset traditional dichotomies on imperial and anti-imperial images, and to the complexity of an iconography merging socialist and imperial visual elements.[296]

In his recollections, Crane explained how he was first 'converted' to socialism by Morris starting from an initial artistic interest. But the circuits of radical London offered a greater variety of views, and in their favourite meeting points one could get acquainted with original people.

> I think this was at Wedde's Hotel in Greek Street, a favourite rendezvous for men and women of advanced political and social opinions of all schools at that time. I met here Pierre Kropotkin, the Russian prince and savant, who had suffered so much for his opinions, and who has won universal respect and sympathy in this country, charming all who have had the pleasure of his acquaintance by his genial manners, his disinterested enthusiasm for the cause of humanity, and his peaceful but earnest propaganda in anarchist-communism, as well as his valuable sociological writings.[297]

Crane also mentioned a New Year's Eve between 1900 and 1901 attended by a group of socialist intellectuals which included the Anarchist Prince.

> The year ended at my friend Charles Rowley's hospitable round table at the National Liberal Club, where he has of late years introduced an agreeable custom of calling a group of his old friends together at a luncheon. His circle included Mr. W. M. Rossetti, Mr. Theodore Watts-Dunton, Mr. Frederick J. Shields, Mr. George Bernard Shaw, Mr. William Rothenstein, Prince Kropotkin, Dr. Richard Garnett, so that in such literary and artistic company it naturally became "a feast of reason and a flow of soul".[298]

A correspondent of Kropotkin,[299] Crane was also a friend of Marsh, with whom he also discussed professional matters in the fields of arts and music: though he never had the ambition of making it a career, Marsh was a violinist and Crane organised evenings with music and theatre where Marsh was invited and consulted for artistic advice.[300] In an 1898 letter to Marsh, Kropotkin referred to the painter in this way: 'I have read from W. Crane a letter saying he is now no more an anarchist but providing a drawing [for *Freedom*] in some time. I hope he will do it.'[301] In his original letter to Kropotkin, Crane had written that he was 'a socialist, and not an anarchist, at present ... Though I think I value individual freedom as much as any man'.[302] This would imply that Crane used to define himself as an anarchist at a certain moment, and this certainly would be new. Anyway, it is clear that Crane represented the tendencies for socialist unity which allowed these intellectual networks, including their anarchist components, to develop across the different ideologies and political affiliations.

The Scottish connection: James Mavor, the Geikies and the Geddeses

After London, the second centre for the interests of Kropotkin and the Recluses in Britain was undoubtedly Edinburgh, for a series of reasons. It was in Edinburgh that Kropotkin disembarked in 1876 at his arrival in Britain, and lived for a few months; however he found life there 'lonely and monotonous',[303] and decided to move to London. The networks and connections between Kropotkin, the Recluses and the Edinburgh entourage of Patrick and Anna Geddes are analysed by an abundant literature,[304] and I will not restate here what is well-known on this connection. However, in this last section I would argue for the importance of Edinburgh and Scotland in general as a place of socialisation for Kropotkin and the Recluses even beyond their connection with the Geddeses. Being mainly rooted in the evolutionism-sensitive scholarly networks of Edinburgh, and in the activist milieus of Glasgow, where anarchism became rather popular thanks to Kropotkin's predication and to the networks of the Socialist League, these connections included famous and less famous names such as Keltie, Herbertson, Robertson-Smith, John Black, John Blackie, Hugh Mill, Archibald and James Geikie, John George Bartholomew and James Mavor. For this purpose, I analyse newly available sources, mainly from the GARF, on the different Scottish collaborations and acquaintances which the Recluses and Kropotkin shared.

As mentioned in Chapter 2, Kropotkin had his first occasion to lecture in Edinburgh and to meet Geddes in 1882. On the occasion of a new trip to Edinburgh in Autumn 1886, Kropotkin possibly had several meetings which inaugurated significant collaborations. According to Woodcock and Avakumović: 'In Edinburgh he stayed with John Stuart Blackie, the classical scholar and translator of Aeschylus, and ... made the acquaintance of a number of men who were later to be among his closest scientific friends. One was James Mavor'.[305] Likewise acquainted with Morris and Carpenter,[306] James Mavor (1854–1925) was a Scottish socialist and was based in Glasgow when he had first the occasion to meet Kropotkin. In 1892, he moved to Toronto, where he taught political economy from 1892 to 1923. Recent scholarship defines Mavor as a forerunner in Canadian geography as his works on Canadian economy gave inspiration to geography, a discipline recently introduced in North American universities. According to John Warketin, Mavor's acquaintance with a number of eminent European geographers contributed to shape his curiosity and his propensity for interdisciplinary approaches. It was the case with Kropotkin, Geddes, Herbertson and Mackinder.[307]

An important number of letters from Mavor to Kropotkin, sent from 1886 to 1916, survive at the GARF, while Kropotkin's answers from 1897 to 1917 survive at the Thomas Fisher Rare Books Library of the University of Toronto. In November 1886, Mavor wrote on behalf of the Scottish Land and Labour League (Scottish section of Morris's Socialist League) to arrange lecture by Kropotkin in Glasgow on 'The revolutionary movement today', on the occasion of his Northern tour.[308] Some weeks later, he wrote again to

acknowledge Kropotkin for sending him copies of *Le Révolté* and to sub-
scribe for the French journal, revealing at the same time his command of
French and his interest in anarchism. In the same letter, Mavor endorsed
Geddes's social projects in Edinburgh, arguing that: 'There is something quite
heroic in his attempt, which cannot fail to have some good effect ... for social
regeneration. With all our revolutionary sympathies, if we can here and now
make the lives even if for a little more bearable, it is our duty to do it, and
therefore I wish the Geddes mission every success.'[309] These remarks well
explain the interest of anarchists such as Kropotkin and the Recluses for
Geddes's relatively 'moderate' essays in social transformation, and matches
the idea of anarchist 'gradualism' by Errico Malatesta in the wider context of
'ethical socialism'. As in other cases, the conversation between Mavor and
Kropotkin soon shifted from politics to scholarship, and vice versa.

Two years later, in his capacity as the editor of the *Scottish Art Review*,
Mavor wrote to the Anarchist Prince requesting him to address a rather
unusual topic for what was his curriculum until that moment.

> I am anxious to get from the most competent hands possible a series of
> articles on the Russian novelists and I should like extremely if it were
> possible for you to give me one on Dostoyevsky ... I want articles also
> upon Tolstoy and Turgenev. I know your opinions about Tolstoy and I
> have suggested Dostoyevsky because I fancy he would be more to your
> mind; I shall be glad to have suggestions from you on the best people to
> do the others.[310]

Kropotkin did not spare his commitment to Mavor's request, as in the
second volume of the journal a paper on Turgenev signed by him appeared as
the first of a series on 'Tourgueneff, Tolstoi, and Dostoievsky'.[311] Dos-
toyevsky's, Turgenev's and Tolstoy's literary works were analysed by Kropotkin in his book on Russian literature published several years later.[312]
Amazingly, this means that, after Knowles, it was again a British editor who
gave Kropotkin an idea for writing papers in periodicals which could be
developed later in a book. The short-lived *Scottish Art Review* published
contributions by Blackie, Carpenter, Crane, Ellis, Geddes, Salt and Arthur
Thomson, confirming that the repeated occurrence of those names in shared
networks was not coincidence, but the marker of a proximity between anarchism
and ethical socialism in artistic, cultural and scholarly work.

In another letter, Mavor showed his interest for the anarchist activity in
London, apologising for having 'missed Mrs. Parsons'[313] and requesting the
address of Russian revolutionary Sergey Stepniak (1851–1895), another of
Kropotkin's friends, then in London. In his following letter, Mavor apologised
again for not having met 'Reclus on that day in London'.[314] Anyway, Mavor's
request for Kropotkin to remember him to Reclus reveals that he had already
met the French scholar. It was accordingly Élie and not Élisée, as some rather
confused notes in Mavor's recollections seem to suggest,[315] while he

definitively met both the Reclus brothers at the occasion of his visit to Paris for the 1900 World fair.[316] Mavor's name must be added to the list of the British political and scholarly acquaintances shared by Kropotkin and the Recluses, and a part of their Scottish connection.

Once in Canada, Mavor continued an abundant correspondence and editorial collaboration with Kropotkin, for instance arranging with Robert Ely the publication of Kropotkin's *Memoirs* in the *Atlantic Monthly*. [317] The collection of Kropotkin's letters held in Canada deserve a specific research work: I would simply say that these long letters which Kropotkin wrote to his Scottish friend further testify the manifold interest, political, scholarly and economic, of Kropotkin's publication activities. In a letter describing how Kropotkin was busy with the *Encyclopaedia Britannica*, likewise a 'Scottish' endeavour at the time of Kropotkin's collaboration with Robertson Smith, the Anarchist Prince argued that: 'Commercially and morally I am bound to do so'.[318] In 1897, Kropotkin attended the Toronto meeting of the British Association for the Advancement of Science, an association where Keltie also played a prominent role.[319] From this journey, the Russian exile wrote a paper on the 'Resources of Canada' for *The Nineteenth Century* with the collaboration of Mavor, who also helped Kropotkin for organising a trip to Vancouver. As Mavor wrote in his recollections: 'An excursion to the Pacific Coast was arranged through the kindness of the Canadian Pacific Railway. A large number of the members took advantage of this excursion. Among these was Prince Kropotkin, who afterwards wrote his impressions in an interesting article in *The Nineteenth Century*'.[320] Kropotkin's stay at Mavor's house in Toronto for a few weeks in 1897 was also an important part of his work for writing *Fields, Factories and Workshops*, as the Scottish economist could help Kropotkin much in studying North-American agricultural and industrial productive systems. In his recollections, Mavor mentioned periodical trips to Britain in which he regularly visited Kropotkin in his Bromley cottage, where he had the occasion to meet Louise Michel and sometimes Tcherkesoff, defined as 'a habitué' in Kropotkin's house.[321] This counters Woodcock's hypotheses on a possible conflict between Tcherkesoff and Kropotkin, also considering that the Georgian Prince dedicated to Kropotkin very admirative pages in his recollections published by Ishill.[322]

On the side of British evolutionist science, it was accordingly in 1882 that Kropotkin met for the first time James Geikie during his first trip to Edinburgh, after Geddes's invitation. This circumstance was evoked by Geikie in a letter he sent to the Anarchist Prince on the occasion of his Scottish tour in November 1886, to organise a meeting at Patrick Geddes's house.[323] In December of the same year, likewise in Edinburgh, Kropotkin seized the occasion for meeting Black on *Britannica* business.[324] In the 1890s, Geikie exchanged several letters with Kropotkin on matters related to his glaciological research in Russia. In 1893, he requested Kropotkin provide a text on the 'Glaciation of Asia' for publishing in the Report of the 63rd Meeting of the British Association for the Advancement of Science. Geikie told

Kropotkin that he was rewriting his own book *The Great Ice Age* and was very interested in Kropotkin's former works on Finland, Siberia and Central Asia, inviting him to submit papers on the 'Glacial Action' and the 'Plasticity of Ice', because 'either the [Edinburgh] Geographical and the Geological Society will be pleased to have your communications'.[325] Geikie played a role in encouraging Kropotkin to translate into English the results of his former physical geography works which will be presented in a series on papers that he published in the *Geographical Journal* from 1893 to 1904. These publications satisfied Geikie's requests, as the British geologist insisted a lot on the fact that most of Kropotkin's former research on glaciations was still available only in Russian, and that he would have been interested in accessing the new findings of Russian explorers and scholars.[326] These papers included orographic maps of Asia which Geikie deemed 'genial'[327] and a series of reviews on ongoing explorations from Russian sources.

In January 1896, Kropotkin travelled again to Edinburgh to lecture. Invited for lunch by Geikie, he took advantage to gather the support of the Scottish scholar for Reclus's Great Globe, an endeavour whose preliminary realisations were then ongoing in Edinburgh by Geddes and by Paul Reclus, refuged there at that time with the pseudonym of George Guyou. Kropotkin related to Keltie the discussion he had with James Geikie, still hiding the name of Reclus's nephew under his pseudonym. What is significant is that Geikie seemed more supportive of Reclus's project than Kropotkin was.

> Geddes has engaged Reclus's young secretary, Mr. Guyou, a very able man, with education [within] engineering work, to begin the relief of Scotland on a 1:100.000 scale, equal scale for heights. Thus, Mr. Guyou has derived a system for making it, the most current and cheapest. James Geikie, with whom I lunched, was delighted of the scheme. What Geddes has not, of course, is money. I advise him to apply to the Newcastle Geogr. Soc., which seems to stay well. What do you think of the scheme? Could or not the Royal GS support it? I am not a great admirer of the equal scales' plan, but Geikie is quite enthusiast of it.[328]

In 1889, as anticipated, it was Robertson-Smith who put Kropotkin in touch with James's brother, Archibald Geikie, who invited the Anarchist Prince to meet him during a sojourn in a hotel close to Kropotkin's house in Harrow on the Hill.[329] Archibald Geikie was likewise one of the authors of reference for Élisée Reclus from the 1860s, when he planned a series of translations of British books for scientific popularisation with the publisher Hetzel, confirming his role of cultural transferor across the Channel, in both directions. Together with Jules Michelet, Geikie appeared to be one of the inspirations of Reclus's *Histoire d'un Ruisseau* and *Histoire d'une Montagne* published by Hetzel, to whom Reclus proposed the translation of Geikie's *Story of a Boulder*, 'which is somehow our book, after ensuring that I am not endeavouring a work which someone else is doing'.[330] Again, publishing market and scientific popularisation were the objects of the exchanges between Kropotkin and Archibald Geikie, who offered the Russian geographer his help for one 'Geographical Series' which he was projecting for MacMillan,[331] even though it seems that this collaboration did not finally take place. Nonetheless, the network of scholars who came from Edinburgh and the surroundings and were mutually acquainted, including Geddes, Keltie, the Geikies, Robertson-Smith and Black for the *Britannica*, and an academic geographer like Herbertson, provided a strong support for both Reclus and Kropotkin confirming the importance of places in the production of knowledge. Edinburgh was not less performant than London as a centre of intellectual interactions and learned societies which fostered scientific sociability.

About the Geddeses, it is worth considering that newly available materials from the Moscow archives further clarify Kropotkin's interest in their project of urban reformation, focusing on its social significance, as it was first directed to workers. In 1887 Anna Geddes, Patrick's wife, exposed their new projects to Kropotkin, such as the new University Hall and its related initiatives for the education of 'the working people of the neighbourhood, [hoping] the students themselves will learn something from living amongst them'.[332] In the following year, Kropotkin wrote to Patrick Geddes to enquire about the progresses of their 'High Street project'[333] and discussed his forthcoming *Nineteenth Century* paper on the integration of labour. These materials also show that Patrick Geddes discussed the theory of mutual aid with Kropotkin, endorsing the Russian geographer in his debates with Huxley on the 'survival of the fittest': 'Still I have many improvements on Darwin ... Huxley does not do better: a little socialism would have [saved him]'.[334] Therefore, the Geddeses were Kropotkin's partners in discussing both activist and scholarly matters and it is very likely that they might have been influenced by his thought on a plurality of topics as several authors, including Lewis Mumford, supposed.[335]

A map-maker who collaborated with Geddes in a project for the construction of a National Institute of Geography in Scotland, John George Bartholomew (1860–1920), was likewise a collaborator of the Recluses and exchanged numerous letters with Élisée Reclus to discuss contributions of the Brussels Geographical Institute for some of Bartholomew's atlases, especially on demographic maps addressing matters of social geography like 'comparative education throughout the world'.[336] Most importantly, an 1895 letter from Geddes to Bartholomew shows that Geddes involved the Scottish mapmaker for 'working Scotland as a type portion of Reclus's Great Globe', and confirms recent scholarship on the direct Reclusian inspiration for Edinburgh's Outlook Tower[337] for its earliest equipment. 'M. Reclus and his favourite pupil M. Guyou ... are elicited in a project in which M. Reclus is much interested, the reconstruction of the building hitherto known as Short's Observatory as a Geographical and Historical Museum'.[338]

The last actor to consider in this Scottish connection is Paul Reclus. After his sojourn at the Cobden-Sanderson's house, Élie's son moved to Edinburg in 1896, and there he worked closely with Geddes. In this sense, Edinburgh was one of the first places for the envisaged production of Reclus's Great Globe. Several letters received by Paul Reclus from different correspondents, including his famous uncle Élisée, survive in the Geddes papers at the National Library of Scotland, though his true name is not always included in the inventories, as his pseudonym George Guyou, often acronymised as 'G.G.' (easy to mistake for some relative of Geddes) might have misled early librarians. This seems to have also misled a great scholar like Martin Miller, who identified Patrick Geddes[339] as the recipient of a letter where Kropotkin recounted in French his 1897 trip to Canada. Indeed, the recipient was Paul Reclus, as revealed without any doubt by Kropotkin's allusion to 'Marguerite [Paul's wife] and the children'.[340] In her seminal work on the French anarchists in London, Constance Bantman mentions Paul Reclus as part of 'the handful of comrades who lived outside London, where there were no informers', attributing to him and Geddes shared interests 'in educational reform which they pursued beyond those years'.[341] In addition, Kropotkin's letters to Marsh confirm that in the mid-1890s Reclus-Guyou was actively collaborating with *Freedom* sending correspondences from Scotland.[342]

Finally, this proximity between the anarchist geographers and this heterogeneous range of 'ethical socialists' became possible thanks to what Ellis called Reclus's and Kropotkin's open-mindedness and 'many-sided nature',[343] a definition which might also match Ellis's personality, as well as those of Carpenter, Salt, Heath, Morris, Mavor, Geddes and many members of these early networks of socialists, scientists, artists and humanists. This chapter has demonstrated that, as Salt also suggested, their humanitarianism should be considered synonymous with humanism rather than with philanthropy, as the 'humane' scholars and activists challenged at the same time anthropocentrism, homophobia, ethnocentrism, authoritarianism and patriarchy. These early battles for civil rights had direct interest for anarchists as far as they remained on the level of cultural production and social engagement, relatively far from governmental politics. Scholarship and cultural work were paramount in this exchange, for instance in Kropotkin's pages for the *Humane Lectures* and those of Reclus for the *Humane Review*. And doubtlessly, for early anarchists, the preferred scholarly discipline was geography.

Notes

1 Ishill, *Peter Kropotkin*.
2 Ishill, *Élisée and Élie*, I.
3 Manton, 'The Fellowship'.
4 Steele, 'Élisée Reclus'.
5 Manton, 'The Fellowship', 283.
6 MacCarthy, *Anarchy and Beauty*, 39.
7 MacCarthy, *Anarchy and Beauty*, 39.
8 MacCarthy, *Anarchy and Beauty*, 43.
9 Manton, 'Ethical socialism'.
10 Heath, 'Agnes Henry'.

11 Morris, *The Collected Letters, vol. III*, 149.
12 Gregory, *Victorians Against the Gallows*, 181.
13 Weinbren, 'Against all cruelty'.
14 Preece, *Animal Sensibility.*
15 Nettlau, 'Peter Kropotkin', 15.
16 Malatesta, 'The most greatly humane man', 38.
17 Levy, 'Foreword'; Turcato, *An Errico Malatesta Reader.*
18 Cresswell, *Geographic Thought.*
19 Reclus, *Correspondance, vol. II*, 221.
20 Reclus, *Correspondance, vol. II*, 222.
21 Reclus, *Correspondance, vol. II*, 86.
22 Reclus, *Correspondance, vol. II*, 87.
23 Reclus, *Correspondance, vol. II*, 87.
24 Reclus, *Correspondance, vol. II*, 221.
25 Papillard, *Michelet.*
26 Reclus, *Correspondance, vol. II*, 223.
27 Reclus, *Correspondance, vol. II*, 221–222.
28 Reclus, *Correspondance, vol. II*, 242–243.
29 Reclus, *Correspondance, vol. II*, 243.
30 Pease, *Richard Heath*, 17.
31 Pease, *Richard Heath*, 16.
32 Reclus, *Correspondance, vol. II*, 243.
33 Pease, *Richard Heath*, 38.
34 Reclus, *Correspondance, vol. II*, 248.
35 Heath, *The English Peasant.*
36 Reclus, *Correspondance, vol. II*, 250.
37 Reclus, *Correspondance, vol. II*, 253.
38 Reclus, *Correspondance, vol. II*, 253.
39 Reclus, *Correspondance, vol. II*, 254.
40 Prichard, *Justice.*
41 Reclus, *Correspondance, vol. II*, 254.
42 Reclus, *Correspondance, vol. II*, 258.
43 London School of Economics Library, Special Collections, 3JBL/26/19, Reclus to Butler, 9 May 1887.
44 Reclus, *Correspondance, vol. II*, 430.
45 Reclus, *Correspondance, vol. III*, 75.
46 Reclus, *Correspondance, vol. II*, 257.
47 Reclus, *Correspondance, vol. II*, 281.
48 Reclus, *Correspondance, vol. II*, 279.
49 Reclus, *Correspondance, vol. II*, 330.
50 Reclus, *Correspondance, vol. III*, 148.
51 Reclus, *Correspondance, vol. III*, 257.
52 Ferretti, 'Evolution and revolution'; White, 'Following the footsteps'.
53 Reclus, *Correspondance, vol. II*, 225.
54 Reclus, *Correspondance, vol. II*, 255.
55 Reclus, *Correspondance, vol. II*, 324.
56 Reclus, *Correspondance, vol. II*, 326.
57 Reclus, *Correspondance, vol. II*, 336.
58 Reclus, *Correspondance, vol. III*, 218.
59 Reclus, *Correspondance, vol. II*, 318.
60 Adams, *Kropotkin.*
61 Reclus, *Correspondance, vol. II*, 322.
62 Reclus, *Correspondance, vol. II*, 325.
63 Reclus, *Correspondance, vol. III*, 234.

64   Reclus, *Correspondance, vol. III*, 235.
65   Heath, 'Élisée Reclus'. 142.
66   Prampolini, *Predica*.
67   Reclus, *Correspondance, vol. II*, 414.
68   GARF, 1129, 2, 2692, Heath to Kropotkin, 15 January 1886.
69   GARF, 1129, 2, 1895, Knowles to Kropotkin, 1882–1906.
70   GARF, 1129, 2, 2692, Heath to Kropotkin, 15 April 1887.
71   Percy William Bunting (1836–1911), then editor of the *Contemporary Review*.
72   GARF, 1129, 2, 2692, Heath to Kropotkin, 28 June 1887.
73   GARF, 1129, 2, 508, 5 letters from Bunting to Kropotkin, 1899–1907.
74   GARF, Fondy P-1129, op. 2 khr 2103, f. 13, É. Reclus to P. Kropotkin, 5 March
      1883; Mečnikov, 'Revolution and evolution'.
75   Ferretti, *L'occidente*.
76   Sellers, 'Our most distinguished'.
77   GARF, 1129, 2, 2692, Heath to Kropotkin, 3 April 1887.
78   Reclus, *Correspondance, vol. II*, 413.
79   Reclus, *Correspondance, vol. II*, 414.
80   GARF, 1129, 2, 2692, Heath to Kropotkin, 11 April 1886.
81   GARF, 1129, 2, 2692, Heath to Kropotkin, 2 May 1886.
82   GARF, 1129, 2, 2692, Heath to Kropotkin, 2 May 1886.
83   GARF, 1129, 2, 2692, Heath to Kropotkin, 2 May 1886.
84   GARF, 1129, 2, 2692, Heath to Kropotkin, 14 May 1886.
85   GARF, 1129, 2, 2692, Heath to Kropotkin, 3 August 1886.
86   GARF, 1129, 2, 2692, Heath to Kropotkin, 11 August 1886.
87   GARF, 1129, 2, 2692, Heath to Kropotkin, 18 January 1887.
88   GARF, 1129, 2, 2692, Heath to Kropotkin, 24 January 1887.
89   Grave, *Mémoires*, 192.
90   Ferretti, 'Organisation'.
91   Pease, *Richard Heath*, 43.
92   Pease, *Richard Heath*, 38–39.
93   GARF, 1129, 2, 2692, Heath to Kropotkin, 28 November 1904.
94   GARF, 1129, 2, 2692, Heath to Kropotkin, 8 July 1899.
95   GARF, 1129, 2, 2692, Heath to Kropotkin, 29 January 1899.
96   GARF, 1129, 2, 2692, Heath to Kropotkin, 18 March 1889.
97   GARF, 1129, 2, 2692, Heath to Kropotkin, January 1889.
98   GARF, 1129, 2, 2692, Heath to Kropotkin, 27 August 1898.
99   GARF, 1129, 2, 2692, Heath to Kropotkin, 9 March 1891.
100   GARF, 1129, 2, 2692, Heath to Kropotkin, 8 May 1892.
101   GARF, 1129, 2, 2692, Heath to Kropotkin, 31 March 1892.
102   GARF, 1129, 2, 2692, Heath to Kropotkin, 17 August 1894.
103   GARF, 1129, 2, 2692, Heath to Kropotkin, 8 May 1892.
104   GARF, 1129, 2, 2692, Heath to Kropotkin, 25 September 1891.
105   GARF, 1129, 2, 2692, Heath to Kropotkin, 1896.
106   GARF, 1129, 2, 2692, Heath to Kropotkin, 5 April 1897.
107   GARF, 1129, 2, 2692, Heath to Kropotkin, 5 July 1899.
108   GARF, 1129, 2, 2692, Heath to Kropotkin, 4 September 1899.
109   GARF, 1129, 2, 2692, Heath to Kropotkin, 24 November 1899.
110   Heath, 'The Kafir', 72.
111   GARF, 1129, 2, 2692, Heath to Kropotkin, 26 June 1899.
112   GARF, 1129, 2, 2692, Heath to Kropotkin, 10 September 1899.
113   GARF, 1129, 2, 2692, Heath to Kropotkin, January 1900.
114   Turcato, *An Errico Malatesta Reader*.
115   GARF, 1129, 2, 2692, Heath to Kropotkin, 15 July 1905.
116   GARF, 1129, 2, 2692, Heath to Kropotkin, 15 July 1905.

117 Eugène Noël (1816–1899), was a French journalist and historian of Normandy, close friend of the Dumesnils. Paris, Bibliothèque Historique de la Ville de Paris, Papiers Dumesnil.
118 GARF, 1129, 2, 2692, Heath to Kropotkin, 25 September 1891.
119 GARF, 1129, 2, 2692, Heath to Kropotkin, 3 April 1899.
120 GARF, 1129, 2, 2692, Heath to Kropotkin, 16 August 1905.
121 GARF, 1129, 2, 2692, Heath to Kropotkin, 16 August 1905.
122 GARF, 1129, 2, 2692, Heath to Kropotkin, 2 May 1898.
123 GARF, 1129, 2, 2692, Heath to Kropotkin, 4 November 1890.
124 GARF, 1129, 2, 2692, Heath to Kropotkin, 5 July 1899.
125 Avakumović and Woodcock, *The Anarchist Prince*, 282.
126 GARF, 1129, 2, 2692, Heath to Kropotkin, 2 March 1902.
127 GARF, 1129, 2, 2692, Heath to Kropotkin, 26 February 1903.
128 GARF, 1129, 2, 2692, Heath to Kropotkin, 5 August 1908.
129 GARF, 1129, 2, 2692, Heath to Kropotkin, 21 June 1903.
130 GARF, 1129, 2, 2692, Heath to Kropotkin, 23 September 1902.
131 GARF, 1129, 2, 2692, Heath to Kropotkin, 19 April 1900.
132 Thompson, *William Morris*, 290.
133 Rowbotham and Weeks, *Socialism and the New Life*, 10.
134 Rowbotham and Weeks, *Socialism and the New Life*, 15.
135 Rowbotham and Weeks, *Socialism and the New Life*, 15.
136 Rowbotham and Weeks, *Socialism and the New Life*, 21.
137 Avakumović and Woodcock, *The Anarchist Prince*, 224.
138 Berry, 'For a dialectic', 1.
139 Rowbotham and Weeks, *Socialism and the New Life*, 53.
140 Rowbotham and Weeks, *Socialism and the New Life*; Tsuzuki, *Edward Carpenter*, 92.
141 GARF, 1129, 2, 1284, Carpenter to Kropotkin, 24 June 1888.
142 GARF, 1129, 2, 1284, Carpenter to Kropotkin, 28 May [no year].
143 GARF, 1129, 2, 1284, Carpenter to Kropotkin, 28 May [no year].
144 GARF, 1129, 3, 200, Carpenter to S. Kropotkin, 27 January 1905.
145 GARF, 1129, 3, 200, Carpenter to S. Kropotkin, 22 August 1916.
146 GARF, 1129, 2, 1284, Carpenter to Kropotkin, 19 March 1917.
147 Rowbotham and Weeks, *Socialism and the New Life*; Tsuzuki, *Edward Carpenter*.
148 'Important letter from Edward Carpenter'. *Freedom*, December 1892, 84.
149 'Important letter from Edward Carpenter'. *Freedom*, December 1892, 85.
150 B. Ward, 'Edward Carpenter'. *Freedom*, September 1929, 3.
151 IISH, Alfred Marsh Collection, 2, Carpenter to Marsh, 9 July 1899.
152 IISH, Alfred Marsh Collection, 3, Carpenter to Marsh, 26 November 1900.
153 Rowbotham and Weeks, *Socialism and the New Life*, 151.
154 'The Wilde case'. *Freedom*, June 1895, 16.
155 E. Carpenter, 'Some recent criminal cases'. *Freedom*, July 1895, 22.
156 E. Carpenter, 'Some recent criminal cases'. *Freedom*, July 1895, 22.
157 E. Carpenter, 'Some recent criminal cases'. *Freedom*, July 1895, 22.
158 E. Carpenter, 'Some recent criminal cases'. *Freedom*, July 1895, 22.
159 Kissack, *Free Comrades*, 101.
160 Carpenter, 'Preface', vii.
161 Cresswell, *Geographic Thought*.
162 Buttimer, *Geography*, 47.
163 Ley, *Geography Without Man*, 14.
164 Carpenter, 'The need', 3.
165 Carpenter, 'The need', 4–5.
166 Kropotkin, *Modern Science*.

167 Carpenter, 'The need', 29.
168 Carpenter, *Civilisation*.
169 Carpenter, 'The need', 27.
170 Carpenter, 'The need', 28.
171 Carpenter, 'The need', 29.
172 Reclus, *HT, vol. 6*.
173 Thomson, 'Humane study', 40.
174 Reclus, *HT, vol. I*, 135.
175 Thomson, 'Humane history', 45.
176 Thomson, 'Humane history', 71.
177 Sheffield City Council, Libraries Archives and Information, Sheffield Archives (hereafter SCC-LAI), Carpenter/Mss/386/345, P. Reclus to Carpenter, 13 February 1929.
178 Carpenter, *My Days*, 2178.
179 'Edward Carpenter'. *Freedom*, September 1914, 69.
180 Rowbotham and Weeks, *Socialism and the New Life*, 147.
181 Rowbotham and Weeks, *Socialism and the New Life*, 154.
182 'The discussion of the sex question'. *Freedom*, July1898, 43.
183 'Free speech for radicals'. *Freedom*, February 1913, 11.
184 'Free speech for radicals'. *Freedom*, February 1913, 11.
185 Hecht, *The End of the Soul*.
186 IFHS, 14 AS 184, Manouvrier to Grave.
187 Ellis, *My Life*, 204.
188 Ellis, *My Life*, 204.
189 Ellis, *My Life*, 205.
190 Ellis, 'Élie Reclus', 48.
191 Ellis, 'Élie Reclus', 49.
192 Steele, 'Élisée Reclus'.
193 Ellis, 'Élie Reclus', 49.
194 Ellis, 'Élie Reclus', 49.
195 Ellis, 'Élie Reclus', 50.
196 GARF, 1129, 2863, Ellis to Kropotkin, 31 October 1896; 9 November 1896; 1 December 1896.
197 Goldman, *Living my Life*.
198 Berry, 'For a dialectic'.
199 Rowbotham and Weeks, *Socialism and the New Life*, 45.
200 Salt H S, 'Alarming condition of the West-End'. *The Commonweal*, 12 June 1886.
201 IISH, Socialist League Archives, 701, Carpenter to the Commonweal manager, 18 December 1885.
202 IISH, Socialist League Archives, 2621, Salt to the Socialist League, 4 January 1887; 1015, Carpenter to Sparling, 31 December 1885.
203 Gregory, *Victorians Against the Gallows*.
204 Clark and Bellamy Foster, 'Henry S. Salt'.
205 Hendrick, *Henry Salt*, 194.
206 Clark, 'Humanitarianism', 106.
207 Tsuzuki, *Edward Carpenter*, 110.
208 Clark and Bellamy Foster, 'Henry S. Salt', 470.
209 Hendrick, *Henry Salt*, 28.
210 Salt, *Seventy Years*, 183–184.
211 Salt, *Seventy Years*, 185.
212 Salt, *Seventy Years*, 185.
213 Salt, 'Access to mountains', 250.
214 Reclus, *Du sentiment de la nature*.

215   Salt, *Seventy years*, 185.
216   H.S. Salt, 'Respectability and its martyrs'. *Freedom*, January 1892, 7.
217   H.S. Salt, 'Respectability and its martyrs'. *Freedom*, January 1892, 7.
218   A[lfred] M[arsh], 'The poet of Freedom'. *Freedom*, July 1896, 91.
219   A[lfred] M[arsh], 'The poet of Freedom'. *Freedom*, July 1896, 91.
220   A[lfred] M[arsh], 'The poet of Freedom'. *Freedom*, July 1896, 91.
221   Ferretti, 'Anarchist geographers and feminism'.
222   'Protest meeting in London'. *Freedom*, February 1897, 15.
223   'The gallows and the lash'. *Freedom*, June 1897, 46.
224   'Clothes and Morality'. *Freedom*, July 1913, 58.
225   Reclus, *HT, vol. VI.*
226   'To our Friends'. *Humanity*, March 1895, 1.
227   Salt, 'Introductory', 1.
228   Salt, 'Introductory', 2.
229   SCC-LAI, Carpenter/Mss 181, M. Clayton to Carpenter, 26 November 1912.
230   Clayton, 'Jottings.
231   Clayton, 'Mary Wollstonecraft'.
232   Fox Bourne, 'The claims', 164.
233   Fox Bourne, 'The claims', 164.
234   Ferretti, 'They have the right'.
235   Davis, *Victorian Holocausts.*
236   Carpenter, 'Empire', 193.
237   Carpenter, 'Empire', 193.
238   Carpenter, 'Empire', 197.
239   Carpenter, 'Empire', 204.
240   Carpenter, 'Empire', 207.
241   Véron, '(Extra)ordinary activism'; White, 'Animal geographies'.
242   White, 'Following in the footsteps', 192
243   Reclus, 'The great kinship', 207.
244   Reclus, 'The great kinship', 208.
245   Reclus, 'The great kinship', 209.
246   Crosby and Reclus, *The Meat Fetish.*
247   Salt, 'Concerning cannibalism', 247.
248   Cobden-Sanderson, 'Domestic economy'.
249   Codello, *La buona educazione.*
250   Heath, 'Francisco Ferrer', 193.
251   Heath, 'Francisco Ferrer', 194.
252   Ferretti, 'Radicalizing pedagogy'.
253   Heath, 'Francisco Ferrer', 195.
254   GARF, 1129, 2, 2209, Salt to Kropotkin, 9 March 1897.
255   Salt, 'The many-sided', 67–68.
256   Salt, 'A poet of socialism', 23.
257   Morris, *The Collected Letters, vol. II*, 453.
258   Morris, *The Collected Letters, vol. II*, 427.
259   Morris, *The Collected Letters, vol. III*, 21.
260   Henderson, *William Morris*, 247.
261   Thompson, *William Morris*, 377.
262   Thompson, *William Morris*, 506.
263   Thompson, *William Morris*, 567.
264   Marshall, *Demanding the Impossible.*
265   Goodway, *Anarchist Seeds.*
266   Kinna, *The Art of Socialism.*
267   MacCarthy, *William Morris*, 29.
268   Freedom Press, *Our First Centenary*, 1986, 10.

269 Wilson, *Anarchist Essays*, 81.
270 Quail, *The Slow Burning Fuse*.
271 Freedom Press, *Our First Centenary*, 1886, 16.
272 'Two lectures'. *Freedom*, January 1893, 4.
273 'William Morris on communism'. *Freedom*, May 1893, 26–27.
274 'In memory of William Morris'. *Freedom*, November 1896, 109.
275 'In memory of William Morris'. *Freedom*, November 1896, 108–109.
276 Edward Carpenter, 'William Morris'. *Freedom*, December 1896, 118.
277 S. Mainwaring, 'Reminiscences of William Morris'. *Freedom*, January 1897, 5.
278 'William Morris and anarchists'. *Freedom*, December 1911, 89.
279 Berneri, *Journey Through Utopia*.
280 P. Kropotkin, 'Edward Bellamy'. *Freedom*, July 1898, 42.
281 Morris, *The Collected Letters, vol. II*, 535.
282 GARF, 1129, 2, 1806 Morris to Kropotkin, 15 September 1888.
283 Bantman, *The French Anarchists*.
284 GARF, 1129, 2, 1806 Morris to Kropotkin, 27 November 1892
285 GARF, 1129, 2, 1806 Morris to Kropotkin, 11 June [no year].
286 GARF, 1129, 2, 1806 Morris to Kropotkin, 7 April [no year].
287 Morris, *The Collected Letters, vol. III*, 265.
288 British Library, RP 6970, Morris to Dryhurst, 6 January 1892.
289 British Library, Morris Papers vol. VIII, MS 45345, Kropotkin to Morris, 10 January 1889.
290 GARF, 1129, 2, 1806 Morris to Kropotkin, 8 November [no year].
291 Avakumović and Woodcock, *The Anarchist Prince*, 220.
292 British Library, Morris Papers, MS 45346, Kropotkin to M. Morris, 25April 1889.
293 Driver, 'In search'.
294 Akermann, *The Imperial Map*.
295 Biltcliffe, 'Walter Crane'.
296 Driver, 'In search'.
297 Crane, *An Artist's Reminiscences*, 255.
298 Crane, *An Artist's Reminiscences*, 480.
299 GARF, 1129, 2, 1447, Crane to Kropotkin, 1892–1913.
300 IISH, Alfred Marsh Papers, 7, Crane to Marsh, 3 May 1897 and 16 May 1897.
301 IISH, Alfred Marsh Papers, 46, Kropotkin to Marsh, 7 July 1898.
302 GARF, 1129, 2, 1447, Crane to Kropotkin, 29 June 1898.
303 Avakumović and Woodcock, *The Anarchist Prince*, 146
304 Chabard, *Exposer la ville*; Meller, *Patrick Geddes*; Ferretti, 'Situated knowledge'; Welter, *Biopolis*. Rich archives on this connection survive in several public libraries, mainly the NLS and the Strathclyde University Library in Glasgow.
305 Avakumović and Woodcock, *The Anarchist Prince*, 210.
306 Morris, *The Collected Letters vol. III*, 466.
307 Warketin, 'James Mavor', 379.
308 GARF, 1129, 2, 1841, Mavor to Kropotkin, 7 November 1886.
309 GARF, 1129, 2, 1841, Mavor to Kropotkin, 2 December 1886.
310 GARF, 1129, 2, 1841, Mavor to Kropotkin, 24 November 1888.
311 Kropotkin, 'Tourgueneff, Tolstoi, and Dostoievsky'.
312 Kropotkin, *Russian Literature*.
313 GARF, 1129, 2, 1841, Mavor to Kropotkin, 3 December 1888.
314 GARF, 1129, 2, 1841, Mavor to Kropotkin, 27 August 1889.
315 Mavor, *My Windows vol. II*, 92.
316 Mavor, *My Windows vol. II, 111*.
317 Mavor, *My Windows vol. II, 93*.

318 Toronto, Thomas Fisher Rare Book Library (TFRB), MS 00119, 10B, Kropotkin to Mavor, 21 February 1901.
319 Withers, *Geography and Science*.
320 Mavor, *My Windows vol. I, 372*.
321 Mavor, *My Windows vol. I,* 93.
322 Tcherkesoff, 'Friend and comrade'.
323 GARF, 1129, 3, 902, J. Geikie to Kropotkin, 30 November 1886.
324 GARF, 1129, 2, 613, Black to Kropotkin, 1 December 1886.
325 GARF, 1129, 3, 902, J. Geikie to Kropotkin, 30 September 1893.
326 GARF, 1129, 3, 902, J. Geikie to Kropotkin, 8 October 1895.
327 GARF, 1129, 3, 902, J. Geikie to Kropotkin, 19 October 1893.
328 RGS-IBG, C7, Kropotkin to Keltie, January 29, 1896.
329 GARF, 1129, 2, 901, A. Geikie to Kropotkin, 3 February 1889.
330 Ferretti, *Lettres*, 68.
331 GARF, 1129, 2, 901, A. Geikie to Kropotkin, 13 February 1889.
332 GARF, 1129, op. 2 khr 824, A. Geddes to Kropotkin, 10 May 1887.
333 NLS, 10524, Kropotkin to P. Geddes, 7 September 1888.
334 GARF, Fondy P-1129, op. 2 khr 895 P. Geddes to Kropotkin, 14 September 1888. Special thanks to Pascale Siegrist for sending her notes on Kropotkin and Geddes.
335 Mumford, *An Introduction*.
336 NLS, Bartholomew Collection, 967, Reclus to Bartholomew, 2 March 1899.
337 Ferretti, 'Situated knowledge'.
338 NLS, Bartholomew Collection, 943, Geddes to Bartholomew, 30 December 1895.
339 Miller, *Kropotkin*, 299.
340 NLS, 10529, Kropotkin to [P. Reclus], 27 November 1897.
341 Bantman, *The French Anarchists*, 55.
342 IISH, Alfred Marsh Collection, 15, Kropotkin to Marsh, 23 April 1903; 16, Kropotkin to Marsh, 26 April 1895.
343 Ellis, 'Élisée Reclus', 49.

6 Conclusion

The relevance of early critical geographies

Between 1914 and 1917, Kropotkin found himself increasingly isolated in British and West-European radical milieus, due to the death of some of his closest friends (Reclus in 1905, Guillaume in 1916) and mainly to the dispute about war, in which his positions were only embraced by a small minority of the international anarchist movement. His experience in 1917–1921 Russia and his encounters with Bolshevik repression are beyond the scope of this work; however, one could assert that Kropotkin's work and legacy can be mostly located in Western Europe, and especially in Britain, rather than in his homeland.

In the case of the Reclus brothers, even though their sojourns in the British Islands were significantly shorter than Kropotkin's, they received a decisive imprinting by their experience of British science and social movements, and a definitive awareness of social injustice by their first-hand witnessing of the miserable conditions of British and Irish workers and peasants. This furthermore demonstrates the importance of places and biographies for the construction of knowledge, and the need for considering its circulations in order to perform contextual readings in the history of science. Anarchism is an especially relevant case in this exercise, given its predominantly transnational nature at the Age of Empire. The processes of cross-breeding between anarchism and geography were likewise located. Although former works addressed the role of Switzerland as the exile land where the first networks of anarchist geographers were established,[1] the role played by the British Isles for the consolidation of these networks was at least threefold.

First, editorial business and the necessity of gathering data and sources were the reasons for the formation of most of the Recluses' networks of acquaintances in that area, being therefore related to their scholarly work. Second, Kropotkin and Élisée Reclus wrote social, political and economic geographies which still inspire radical scholarship, in works such as *Fields, Factories and Workshops* and Reclus's chapter on British imperialism in *L'Homme et la Terre*, based on their experiences in the British Isles. Third, in Britain intellectual anarchists found sufficient room for having intellectual exchanges with a very heterogeneous range of scholars and activists, from those who were a far cry from anarchism like Mackinder, to those who were very akin to their ideas, such as Morris, Cowen, Geddes and the 'humanitarians'. Their respective British and Irish milieus and experiences informed (in different ways) Reclus's and Kropotkin's thinking on social questions, on science and evolutionism, on internal and external colonialism. For elaborating the concept of mutual aid, the role played by the intellectual milieus of the British Isles as the cradle of evolutionist science were likewise paramount. If Russian traditions were influential in inspiring cooperative ideas of evolution, Kropotkin's conversations with British evolutionists who were not aligned with Huxley's positions show that cooperative understandings of Darwinism were not extraneous to British scholarship.

However, Kropotkin and the Recluses acted as cultural transferors between the British Isles and the Continent in both directions. Their ideas had a wide impact and a specific reception in Britain and Ireland, thanks to the huge English-speaking readership of Kropotkin (and partially of the Recluses) and to their acquaintances, who often contributed to the indirect spread of their ideas through their own works and networks. This was the case with the heterogeneous British and Irish circuits where both activist and scholarly aspects of Reclus's and Kropotkin's geographies found a fertile ground for discussing and publishing on a range of subjects such as anti-colonialism, feminism, defence of the environment and animal rights. Kropotkin and the Recluses played these roles of cultural transferors thanks also to their multilingual and multicultural skills, which can surely be a good basis for opening a debate on the problems of multilingualism and cosmopolitism in today's geographical (and anarchist) scholarship.

From the standpoint of historical geographies, this work sheds new light on forgotten aspects of the geographical tradition, namely its 'alternative' tendencies, and provides insights for today's critical scholarship engaging with the return of theories such as Creationism, Malthusianism and environmental determinisms analysed by recent geographers' works.[2] Methodologically, this book has shown how discourse analyses alone are not sufficient in order to well understand authors and their scholarly and activist networks: cultural and historical contexts, as well as the spatialities and mobilities of knowledge, need to be addressed. Together with the other elements analysed to explain the welcoming of the anarchist geographers in the British Isles, such as liberalism and editorial business, the tolerance towards exiles and refugees which existed there can also be attributed to some traditional British isolationism, as discussed by Reclus in *L'Homme et la Terre*. This is not to undermine the importance of the British soil as a haven for several kinds of refugees: if some cultural isolationism constituted a limitation on the standpoint of transcultural dialogue, this was very practical for the exiles, who were not persecuted in Britain for whatever problem they had with continental tribunals and police forces.

From the standpoint of print cultures, this work allows a reassessment of the role that publishing business played for the bargains established between the anarchist geographers and their editors and fellows. Now, after my discussion, their importance stands clearer and clearer, confirming the effectiveness of tracing editorial networks to understand the construction of geographical (and generally scholarly) knowledge at a time where geography was not yet established as an academic discipline in the British Isles and in most of European countries. The context of expanding readerships and literacy for both books and periodicals, and the related technical innovations were key to understanding those bargains. However, I would argue that, for some aspects, my survey of *Freedom* challenges commonplace distinctions between 'scientific journalism' and 'political journalism',[3] distinctions which were generally taken for granted by much traditional anarchist historiography. This allows the highlighting of the coherence of anarchist geographers' efforts toward publication in different supports. Whereas substantial materials for some key Kropotkin's anarchist books were first published in *The Nineteenth Century*, others, like parts of *The Conquest of Bread*, were first published in *Freedom*. As anticipated in the introduction, this work also analyses a case where books and periodicals were complementary to these strategies of maximizing dissemination, contributing to show that there is not necessarily a rigid distinction between geographies of the book and geographies of periodical publications.

From a more political standpoint, my survey of *Freedom, Humane Review* and the related publications and unpublished correspondences shows the early commitment of anarchists, and especially the more intellectual activists engaging with geography, to what today would be called intersectional or multiple-axes approaches. Though an important part of their activities focused on the more traditional field of class struggle, it is now impossible to argue that early anarchists focused on single-axis thinking, as they paid equal attention to topics such as workers' rights, anti-colonialism (including the Irish case), anti-racism, women's rights, issues on sex and 'homogenic love', civil rights and liberties, issues on capital punishment and prisoners' treatment, protection of the environment, antimilitarism, vegetarianism, interspecies solidarity and animal rights. The famous anecdote about an argument between Emma Goldman and Kropotkin one day about the political relevance of sex, when the Anarchist Prince finally admitted that his interlocutor was probably right in arguing ironically that he was too old to fully appreciate the problem,[4] only confirms the richness and pluralism of these debates in the anarchist field, manifestly richer than in any other socialist movement at that time. The study of the specific relations between anarchism and British 'ethical socialism' has shown that humanitarianism and humanism are not necessarily synonymous with speciesism and anthropocentrism, and not even with paternalism or charity. On the contrary, the humanistic aspects of the anarchist traditions exposed in this work and in recent Malatesta scholarship by Levy, Turcato and others, can nourish present-day ideas on intersectionality, as well as the current rediscovery of humanistic approaches in radical and anarchist geographical scholarship.[5]

Likewise, the study of these networks exposes the oddity of the contemporary fashion of attributing to different forms of anarchist activism definitions such as 'Big A' and 'small a'. With all my respect for post-structuralist contributions, I would contend that complexity, diversity, fluidity, porosity and plural belongings have always characterised anarchism from the nineteenth century onward. As the journal *Freedom*'s networks demonstrate, the weight and influence of anarchism must be assessed considering its pluralist networks and the wider sympathies and synergies that this idea always had, well beyond the cores of self-declared and fully fledged anarchist militants. As discussed throughout this book, these wider interactions were eventually facilitated by the complex and pluralistic traditions of British socialism and its multiple legacies, including the revolutions of the seventeenth century and the Dissent, and doubtlessly by the anarchist engagement in scholarship and people's education, primarily through geography. Finally, authors like Adams started to investigate the reception and long-lasting influence of the anarchist geographers (eventually Kropotkin, with special reference to Herbert Read)[6] in Britain in more recent times: more work should be done in this direction to further extend these invaluable contributions.

Notes

1 Ferretti, *Élisée Reclus*.
2 Livingstone, 'Changing climate'; Mayhew, *Malthus*.
3 Walter and Becker, 'Introduction', 7
4 Avakumović and Woodcock, *The Anarchist Prince*, 257–258.
5 Springer, 'Earth writing'.
6 Adams, *Kropotkin*.

Appendix A

Pyotr Kropotkin, 'Natural selection and mutual aid'

Humane Science Lectures, *1897, 182–186 [transcribed by Edward Carpenter on attendants' notes]*

This lecture, which was the second in the series, was not written, and it has been impossible to reproduce it except in the following much epitomized form, which was printed at the time in *Humanity*. After remarking that the subject of mutual aid is essential to any philosophy of humane science, the lecturer pointed out that, in the continual development of science, periods necessarily occur—as now—when there seems to be no satisfactory progress, there being a temporary pause while preparation is made for a new step, an advance to further generalisations. Untrained minds, impatient at the delay, attempt to supply more than science can give; and this, together with the patronage of Church and State, tends to impair the usefulness of Science. Political Economists who know nothing of the life of the people and the actual conditions of production, write learned works which are accepted as scientific; and in the same way Natural History is studied in closed laboratories and not, as Audubon studied it, in the open forests. Thus, conclusions are arrived at which are antagonistic to human nature, and it is believed that science is somehow instructing us to take each other by the throat. But science has no such prescription for us, and indeed no prescription at all; it merely tells us facts—what consequences follow what causes. 'Darwinism' is now-a-days made to answer for every sort of outrage, is the explanation of every villainy, as, for example in our recent treatment of the Matabele, whose extinction is justified on the plea that 'black men must go', 'it seems cruel, but it is their inevitable destiny', and other equally 'scientific' assertions. Nature, according to Huxley's theory, is no better than a gladiatorial show, where each being is against each, and there is no need for the spectators to turn their thumbs downwards (the signal for the *coup de grâce*), because no quarter is ever shown in any case, since life is a continual free fight. But, said the lecturer, Darwin does not teach this. He proves that there is a struggle for existence, in order to put a check on the inordinate increase of species. But this 'struggle' is not to be understood in a crude and exclusive sense; there is a law of competition, but there is also what is still more important—a law of mutual aid, and as soon as the scientist leaves his laboratory and comes out into the open woods and meadows, he sees the importance of this law. Only those animals who are mutually helpful are really fitted to survive; it is not the strong, but the co-operative species that endure.

Instances of mutual aid, of which any number might be quoted, may be seen even in the less developed forms of life. Land-crabs migrate in columns from sea to land; and the lecturer narrated how he had watched an over turned king-crab at the Brighton Aquarium laboriously set on its feet again by the repeated efforts of its companions. The good will of ants is signified by a free gift of food from full crop to empty crop, and this pact of friendship is not confined to individuals but extends to whole nests, thus showing that the Stomach exists not for individuals only but for the community. Natural Selection comes to aid those species that are social. Much is said of birds and beasts 'of prey.' But birds of prey are comparatively few in number, whereas the other kinds, where man has not come on the scene, are countless, as for example, the passenger pigeons in America, which once flew in such flocks as to obscure the sun for days, or the various species which in high northern latitudes breed in immense numbers and all co-operate to scare away the intruding robber. So, too, with the mammals. There is much talk of the savagery of lions and tigers, but how few they are by comparison, let us say, with the whole villages of prairie-dogs, who live in perfect amity and comradeship! The lecturer further instanced the vast processions of buffaloes that might once be seen in North America, the beasts of prey that followed them being merely the scavengers of nature. The highest form of association among animals is of course to be seen among Monkeys, whose combined defence is so perfect that it has been said that they seldom die any but a natural death, and instances are recorded of their carrying off the dead body of a comrade from the tent of his human murderer. Mutual aid is thus a very substantial element in existence, and not for utilitarian purposes only, but for the simple enjoyment of life. The highest developed in every class is the most sociable, because the increased length of years which association secures is favourable to the increase of Experience.

It remains to apply this principle to human science. 'It may be true of the animals,' it is said, 'but is it true of man? Is it true of savages?' a doubt to which even Spencer and Huxley have in some degree lent their sanction. But those who have lived among savages know that it is true. The records of the early travellers in Oceania and the Pacific Isles led to that conception of an ideal 'state of nature' on which so much ridicule has been poured by later writers; but, as a matter of fact, scientific investigation has revealed in these races a remarkable wealth of institutions for mutual aid, and the existence of a happy and peaceful society without authority or government. In the tribal state which preceded the family every possession was in common, and whatever was held by the individual returned to the tribe at his death. In the village communities of so called 'barbarians', there was a common ownership of land, and a jury system which settled quarrels by arbitration—intelligence having been developed to this extent out of mutual aid.

In spite of the teachings of supposed scientific authorities, mutual aid exists largely among the poorer classes of to-day; and if we leave printed matter, and go to study the actual facts of life, we find great material to support this belief. It was because Huxley over-looked this law of mutual aid, that he was driven to look for help from another quarter, and so gave some countenance to the idea of a return to supernaturalism. The process of Mutual Aid has been developed from the first, through countless ages, among animals, and its application to Man is only a continuance of the same law. Let us note the lesson of Nature. In times of scarcity, how do animals and birds act? They migrate; or, like the ants, take concerted measures to provide themselves with food. Yet Man, the highest of animals, thinks he has no option but to rob his fellows, as Englishmen have robbed and spoiled the Matabele. There is no need of any extraneous or super natural help or admonition. All the elements of morality are inherent in Nature, if we would but study them.

Appendix B

Élisée Reclus, 'War', **Freedom,** *May 1898*

War is upon us, terrible war with its atrocities and unspeakable stupidity ... We hear its distant echo. Each one of us has friends or relations calling themselves heroes because they massacre Matabeles or Malgays, Dervishes or Dacoits, inhabitants of the Philippine Isles or Cuba, coloured men, whites or blacks. But danger breaks forth around as, it already presses on us. The Spaniards, our neighbours, and civilised English-speaking men—North Americans, rush on one another with cries of hatred, coarse words and deadly weapons. An explosion of hatred and fury precedes the cannon's roar and the bombardment of cities. The American government appeals to Edison's genius, to the science of all inventors in order that they discover new wonders in the art of exterminating their fellow creatures. Yet our engineers are skilful and may be congratulated on the success of their works of death. For a devil who dreams of universal destruction, it must be a fine sight to see a volley suddenly transform a whole battalion of able-bodied young men into a quivering mass of flesh torn by *dum-dum* bullets! It must also be a refined pleasure for an enemy of mankind to know that the mere contact of two wires produces a shock like a thunderbolt and blows a whole district of a town, with stones, bricks, brains and limbs mixed up, into space. But this does not suffice them. What all these warlike people want now is to cause the total destruction of man and of his works in an immense space, it is to outdo the blind power of nature by the conscient ferocity of man. They must surpass earthquakes and hurricanes, destroy not only haphazard but by an infallible method.

'No 'false sentimentality!' they say to us. 'War is war!' Well then, so be it. It is because war brings all these terrible consequences in its train that it behoves us to grasp it in its origin and to stop all manifestations from the beginning. A viper must be crushed in the egg. What is there astonishing in war breaking ... the whole of society, as constitutions and institutions have created it, rests on hostile interests and that a rumbling struggle rages in the depths? Pent up hatred must burst forth from time to time, and we should show bad grace in complaining if we ourselves contributed to augment this hatred, or if inert and cowardly before violence and injustice were complacently let it accumulate day by day!

It would therefore be trouble thrown away to cry out against Spain and the United States. When the masses are lashed into an unreasoning fury it is useless to add your voice to the uproar. It is vanity to preach morality to men drunk with alcohol or blood. What is the use of telling them they are exploitable material for spirit merchant, for monopolisers of corn, spirits and petroleum?

But we who are not made drunk by war, are we not guilty? Have we protested with all our energy, have we rebelled against initial abominations that lead to war, do we always make a stand against injustice, even if it is committed by our countrymen? Now, these social crimes press on us, encompass us on all sides. Here in the British Isles we have the horrors of the workhouse and of the tread-mill, the great prison of Ireland, and our rich folk live on the spoils of half the world. In France, we have committed the crime of our colonial conquests, the shame of our magistracy ready for anything, the incubus of our army, organised against the people, with its stuff brought up in the school of Jesuitism, with its officers that sentence privates guilty of a gesture to death or to the horrors of Biribi, while they adulate traitors of high degree. And the Italian Government? We let them shoot down starvelings, and put those who ask for bread for the poor into prison. As to Spanish ministers, who now drive their people to butchery, they profit by excitement and disturbances, and forget to liberate the men they have tortured. And yonder, on the other side of the Atlantic, American judges (of whom the public lazily becomes the accomplice) acquit an employer who has caused workmen to be massacred by the score!

Do not let us forget these atrocities. May war, with its terrible upheavals not let us lose sight of social injustice, that monstrous inequality among men which is the primary cause of all strife.

Appendix C

[Edward Carpenter], To Peter Kropotkin from Friends in Great Britain and Ireland, 1912

Sheffield City Council, Libraries Archives and Information, Sheffield Archives, Carpenter/Mss 181

It is with great pleasure that on this occasion of your seventieth birthday, we take the opportunity of expressing our true admiration of your life-work and the noble elements of character which have directed and inspired it. Your services in the cause of Physical Science, your contributions to Geographical and Geological discovery, your criticisms and extensions of the Darwinian theory of Evolution are matters of world-wide recognition, and have greatly enlarged our understanding of general Nature—while your criticisms and emendations of current Political Economy have similarly widened our outlook on human and social life.

You have taught us to rely in social life on that most important force, the voluntary principle, which has inspired so much of the best life in all ages of the world, and which is now among the modern societies taking its place as the leading factor in their development—in contradistinction to the merely regulative and governmental principle, which in the form of over-legislation certainly tends to render a people deficient in originality and initiative. The entirely natural growth and expression of this voluntary co-operation in the life of mankind you have abundantly illustrated for us from your studies of the animal world.

Nor, needless to say, has your work in these directions been merely theoretical. On the contrary, a rare devotion to the cause of the suffering and oppressed in all lands has led you on many occasions to risk your life and to endure times of extreme hardship and persecution—an example which we cannot forget and which cannot fail to affect us profoundly. We congratulate you indeed upon the splendid record of work done in this and other directions; and at the same time, we congratulate ourselves that we have you living among us and honouring our land by making it your home. May it long continue thus to be your home—a pleasant dwelling-place for you in the hearts of your friends, and a secure vantage-ground from which to carry on your work.

With every salutation and good wish, from these your friends:

Norman Angell; H. Granville Barker; Samuel A. Barnett; Arnold Bennett; Rutland Boughton; H.N. Brailsford; Jane E.M. Brailsford; Mary S. Burt; W.P. Byles; S.A. Byles; George Cadbury; R.J. Campbell; Alfred Carpenter; Edward Carpenter; G.K. Chesterton; G.B. Clark; Joseph Clayton; Frederick C. Conybeare; Walter Crane; Charlotte Despard; Havelock Ellis; Elizabeth H. Ford; Isabella O. Ford; John Galsworthy; A.G. Gardiner; Constance Garnett; Edward Garnett; Patrick Geddes; G.P. Gooch; R.B. Cunninghame Graham; Alice Stopford Green; J. Frederick Green; J. Keir Hardie; Beatrice Harraden; Stewart D. Headlam; William Heinemann; Mary St. Helier; John A. Hobson; Henry Holiday; Ford Madox Hueffer; Violet Hunt Hueffer; Holbrook Jackson; J. Scott Keltie; George Lansbury; Oliver Lodge; Robert Lynd; Sylvia Lynd; J.A. Murray MacDonald; Haldane MacFall; John Masefield; H.W. Massingham; Aylmer Maude; C.E. Maurice; Emily Southwood Maurice; Hugh Robert Mill; May Morris; Felix Moscheles; Henry W. Nevinson; James O'Grady; Constance Phillott; Eden Phillpotts; Caroline E. Playne; Horace Plunkett; Arthur Ponsonby; S.K. Ratcliffe; Katie M. Ratcliffe; H.D. Rawnsley; W.B. Richmond; Charles Rowley; Seebohm Rowntree Russell; George W. Russell; John Russell; Rollo Russell; Henry S. Salt; Anne Cobden Sanderson; C.P. Scott; G. Bernard Shaw; Upton Sinclair; G.M. Trevelyan; J. Fisher Unwin; Jane Cobden Unwin; Raymond Unwin; Paul Vinogradoff; Emery Walker; Josiah C. Wedgwood; H.G. Wells; Richard Whiteing; Gaylord Wilshire; Charlotte M. Wilson; Norbert Wylie; Israel Zangwill.

Archives

Canada

University of Toronto, Thomas Fisher Rare Book Library (TFRB), James Mavor Collection

Ireland

Dublin – National Library of Ireland (NLI), Department of Manuscripts
Nannie Dryhurst Family Papers
Roger Casement Papers

United Kingdom

London – Royal Geographical Society with Institute of British Geographers (RGS-IBG), Manuscripts Department.
London – British Library (BL), Departments of Manuscripts
London – Anthropological Institute of Great Britain and Ireland Archives
London – London School of Economics (LSE) – Special collections
London –City of Westminster Archives Centre (CWAC), Knowles Papers
Cambridge – Darwin Correspondence Project http://www.darwinproject.ac.uk/
Sheffield – Sheffield City Archives, Edward Carpenter collection.
Edinburgh, National Library of Scotland (NLS)
Department of Manuscripts, Geddes Papers.
Department of Maps, Bartholomew Archive
Glasgow, Strathclyde University, Special Collections, Patrick Geddes Archive

Netherlands

Amsterdam – International Institute of Social History (IISH), Archives

France

Pierrefitte-sur-Seine – Centre d'Accueil et de Recherche des Archives Natio-nales – Institut Français d'Histoire Sociale (IFHS)

- 14 AS 232, Dossiers Élisée Reclus
- 14 AS 184b, Lettres à Jean Grave

Paris, Bibliothèque Historique de la Ville de Paris (BHVP), Papiers Dumesnil

Switzerland

Neuchâtel – Bibliothèque Publique et Universitaire (BPUN), Ms 1991/10

Russia

Moscow – Gosudarstvennyi Arkhiv Rossiiskoi Federatsii (GARF)
Fondy P-1129, Pëtr Aleksejevič Kropotkin, op. 2–3

Acronyms

NGU – Reclus, E., *Nouvelle Géographie universelle*, Paris, Hachette, 1876–1894, 19 vols [quoted with acronym and volume number in the endnotes]
HT – Reclus, E., *L'Homme et la Terre*, Paris, Librairie Universelle, 1905–1908, 6 vols [quoted with acronym and volume number in the endnotes]

Newspapers

Articles from British newspapers [*Pall Mall Gazette; The Morning Advertiser; The Berkshire Chronicle; Cambridge Independent Press; The Glasgow Herald; The Yorkshire Post and Leeds Intelligencer; The Greenock Telegraph and Clyde Shipping Gazette; Newcastle Courant; London Evening Standard; Illustrated Times*] and newspaper-like political journals such as *Freedom, Commonweal, Humanity* and *Temps Nouveaux* are quoted only in the endnotes and not in the final bibliography.

Freedom, a journal of anarchist communism, whose collections from 1886 to 1917 have been integrally consulted for this research, is mentioned simply as *Freedom* in the text and in the notes.

Printed sources such as books and journals have been consulted in different libraries, including the Department of Early Printed Books at the Trinity College Library in Dublin, the British Library in London and the library of the International Institute of Social History in Amsterdam.

Quotations

All quotations from archives, secondary sources and literature originally in languages other than English have been translated by the author. In the case of
quotations from works and unpublished letters written in English by non-native
speakers such as Kropotkin and Reclus, texts which contain a certain number of
mistakes and non-idiomatic locutions, I have maintained the original grammar
and syntax, while editing their spelling to align it with current uses.

References

Adams, M.S., *Kropotkin, Read and the Intellectual History of British Anarchism*. London: Palgrave MacMillan, 2015.

Adams, M.S. and R. Kinna (eds), *Anarchism, 1914–18: Internationalism, Anti-militarism and War*. Manchester: Manchester University Press, 2017.

Agulhon, M., *La sociabilité méridionale, Confréries et associations dans la vie collective en Provence orientale la fin du XVIIIe siècle*. Aix-en-Provence: La pensée universitaire, 1966.

Akermann, J. (ed.), *The Imperial Map, Cartography and the Mastery of Empire*. Chicago: University of Chicago Press, 2009.

Alavoine-Muller, S. 'Un globe terrestre pour l'Exposition universelle de 1900. L'utopie géographique d'Élisée Reclus'. *L'Espace Géographique* 32, no. 2(2003): 156–170.

Alavoine-Muller, S. 2007. 'Introduction'. In Reclus, É., *Les États-Unis et la guerre de sécession: articles publiés dans la Revue des Deux Mondes*, 1–70. Paris: Editions du CTHS.

Aldred, G.A., *Dogmas Discarded*. London: The Bakunin Press, 1913.

Aldred, G.A., *Dogmas Discarded, Part Two, an Autobiography of Thought, 1902–1908*. Glasgow: The Strickland Press, 1940.

Allen, J., *Joseph Cowen and Popular Radicalism on Tyneside, 1829–1900*. London: Merlin, 2007.

Altena, B., 'English and Dutch anarchists and the South-African War, 1899–1902'. Paper presented at the ESSHC 2016, Valencia.

Anderson, B., *Under Three Flags: Anarchism and the Anti-Colonial Imagination*. London: Verso, 2007.

Anderson, B., *Life beyond Boundaries*. London: Verso, 2015.

Anthropological Institute, 'President's Address'. *The Journal of the Anthropological Institute of Great Britain and Ireland*, no. 8(1879): 402–425.

Arrault, J.B., 'À propos du concept de Méditerranée: expérience géographique du monde et mondialisation'. *Cybergeo*, 2006, http://www.cybergeo.eu/index13093html

Ashmore, P., R. Craggs, and H. Neate, 'Working-with: talking and sorting in personal archives'. *Journal of Historical Geography* 38(2012): 81–89.

Avakumovič, I. and G. Woodcock, *The Anarchist Prince: A Biographical Study of Peter Kropotkin*. New York: Shocken Books, 1971.

Avrich, P., *The Russian Anarchists*. Princeton: Princeton University Press, 1967.

Baigent, E. 'The geography of biography, the biography of geography: rewriting the *Dictionary of National Biography*', *Journal of Historical Geography*, no. 30(2004), 531–551.

Baines, D. and R. Woods, 'Population and regional development'. In R.C. Floud and P. Johnson (eds), *The Cambridge Economic History of Modern Britain, Volume II: Economic Maturity, 1860–1939*, 25–55. Cambridge: Cambridge University Press, 2004.

Balinisteanu, T., *Violence, Narrative and Myth in Joyce and Yeats: Subjective Identity and Anarcho-Syndicalist Traditions*. Basingstoke: Palgrave Macmillan, 2012.

Bantman, C., *The French Anarchists in London, 1880–1914: Exile and Transnationalism in the First Globalisation*. Liverpool: Liverpool University Press, 2013.

Bantman, C., 'Louise Michel's London years: a political reassessment (1890–1905)', *Women's History Review* (2017), doi:10.1080/09612025.2017.1294393

Bantman, C., 'Jean Grave and French Anarchism: A Relational Approach (1870s–1914),' *IRSH* 62(2017), 451–477. doi:10.1017/S0020859017000347

Bantman, C. and B. Altena, *Reassessing the Transnational Turn: Scales of Analysis in Anarchist and Syndicalist Studies*. New York: Routledge, 2015.

Barclay, H., *People without Government: An Anthropology of Anarchy*. London: Kahn and Averill, 1996.

Baring, E., 'Ideas on the move: context in transnational intellectual history'. *Journal of the History of Ideas* 77, no. 4(2016): 567–587.

Bassin, M., *Imperial Visions: Nationalist Imagination and Geographical Expansion in the Russian Far East, 1840–1865*. Cambridge: Cambridge University Press, 1999.

Baubérot, A. and F. Bourillon, *Urbaphobie, la détestation de la ville aux 19e et 20e siècles*. Pompignac: Éditions Bière, 2007.

Becker, H., 'Notes on *Freedom* and the Freedom Press 1886–1928'. *The Raven*, no. 1 (1986): 4–24.

Berneri, M.L., *Journey Through Utopia*. London: Freedom Press, 1952.

Berry, D. 'For a dialectic of homosexuality and revolution', Conference on Socialism and Sexuality. Past and Present of Radical Sexual Politics, Amsterdam, 3–4 October 2003, https://theanarchistlibrary.org/library/david-berry-for-a-dialectic-of-homosexuality-and-revolution

Bey, M.F., 'Egypt's demand'. In N.F. Dryhurst (ed.), *Nationalities and Subject Races*, 11–19. Westminster: King & Son, 1911.

Biltcliffe, P., 'Walter Crane and the Imperial Federation Map showing the extent of the British Empire (1886)'. *Imago Mundi* 57, no. 1(2005): 63–69.

Blunt, A. and J. Wills, *Dissident Geographies: An Introduction to Radical Ideas and Practice*. London: Routledge, 2000.

Boardman, P. *Patrick Geddes: Maker of the Future*. Chapel Hill: University of North Carolina Press, 1944.

Bond, D., 'Plagiarists, enthusiasts and periodical geography: A.F. Büsching and the making of geographical print culture in the German Enlightenment, c.1750–1800'. *Transactions of the Institute of the British Geographers*, no. 42(2017): 58–71. doi:10.1111/tran.12153

Bond, D., 'Enlightenment geography in the study: A.F. Büsching, J.D. Michaelis and the place of geographical knowledge in the Royal Danish Expedition to Arabia, 1761–1767'. *Journal of Historical Geography*, no. 51(2016): 64–75. doi:10.1016/j.jhg.2015.09.002

Brun, C., 'Introduction'. In C. Brun (ed.), *Élisée Reclus: Les Grands Textes*, 19–52. Paris: Flammarion, 2014.

Brun, C., 'Élisée Reclus, une chronologie familiale'. *Raforum*, (2015) http://raforum.info/reclus/spip.php?article455

Buttimer, A., *Geography and the Human Spirit*. Baltimore: Johns Hopkins University Press, 1993.

Cahm, C., *Kropotkin and the rise of revolutionary anarchism: 1872–1886*. Cambridge: Cambridge University Press, 1989.

Carpenter, E., 'Preface'. In E. Carpenter (ed.), *Humane Science Lectures*, vii–viii. London: George Bell & Sons, 1897.

Carpenter, E., 'The need of a rational and human science'. In E. Carpenter (ed.), *Humane Science Lectures*, 1–34. London: George Bell & Sons, 1897.

Carpenter, E., 'Empire, in India and elsewhere', *The Humane Review*, no. 1(1900): 193–207.

Carpenter, E., *My Days and Dreams*. London: George Allen & Unwin, 1916

Carpenter, E., *Civilisation, its Cause and Cure, and Other Essays*. London: Allen & Unwin, 1921.

Carroll, C. and P. King, *Ireland and Postcolonial Theory*. Cork: Cork University Press, 2003.

Chabard, P., 'Un éléphant blanc dans le Périgord Noir: la Tour de Paul Reclus et Patrick Geddes à Domme'. *Les Cahiers Élisée Reclus*, 53(2005): 1–2.

Chabard, P., 'Towers and globes: architectural and epistemological differences between Patrick Geddes's Outlook Towers and Paul Otlet's Mundaneums'. In W. Boyd Rayward (ed.), *European Modernism and the Information Society, Informing the Present, Understanding the Past*, 105–125. London: Routledge, 2008.

Chabard, P., *Exposer la ville: Patrick Geddes (1854–1832) et le Town Planning movement*. Unpublished PhD thesis, University of Paris, 2008.

Cho, S., Crenshaw, K.W. and L. McCall, 'Toward a field of intersectionality studies: theory, applications, and praxis'. *Signs: Journal of Women in Culture and Society* 38, no. 4(2013): 785–810.

Clark, A., 'Humanitarianism, Human Rights, and Biopolitics in the British Empire, 1890–1902'. *Britain and the World* 9, no. 1(2016): 96–115.

Clark, B. and J. Bellamy Foster, 'Henry S. Salt, socialist animal rights activist: An introduction to Salt's *A Lover of Animals*'. *Organization & Environment* 13, no. 4 (2000): 468–473.

Clark, J.P. and C. Martin. *Anarchy, Geography, Modernity: Selected Writings of Élisée Reclus*. Oakland: PM Press, 2013.

Clayton, M., 'Jottings in jail'. *The Humane Review*, no. 8(1907): 157–166.

Clayton, M., 'Mary Wollstonecraft and woman's enfranchisement'. *The Humane Review*, 10(1910), 205–218.

Cobden-Sanderson, A., 'Domestic economy and the ideal home'. *The Humane Review*, no. 2(1901): 125–134.

Cobden-Sanderson, A., 'Élie and Élisée Reclus'. In J. Ishill (ed.), *The Brothers Élie and Élisée Reclus*, 43–46. Berkeley Heights: Oriole Press, 1927.

Codello, F., *La buona educazione*. Milano: Angeli, 2005.

Craggs, R. and C. Wintle, *Cultures of Decolonisation, Transnational Productions and Practices, 1945–70*. Manchester: Manchester University Press, 2016.

Crane, W., *An Artist's Reminiscences*. New York: MacMillan, 1907.

Cresswell, T., *Geographic Thought: A Critical Introduction*. Chichester: Wiley-Blackwell, 2013.

Crosby, E. and E. Reclus. *The Meat Fetish: Two Essays on Vegetarianism*. London: Humanitarian League, 1905.

Davies, A., 'Exile in the homeland? Anti-colonialism, subaltern geographies and the politics of friendship in early twentieth century Pondicherry, India'. *Environment and Planning D, Society and space* 35, no. 3(2017): 457–474.

Davis, M., *Victorian Holocausts.* London: Verso, 2001.

De Claparède, R., *L'évolution d'un État philanthropique: les origines de l'État indépendant du Congo.* Geneva: Atar, 1909.

De Rousiers, P., 'Études de géographie sociale: la Géographie Universelle d'Élisée Reclus'. *La Reforme Sociale,* no. 8(1884): 511–517.

Diamond, J., *Guns, Germs and Steel.* London: Cape, 1997

Dicken, P., *Global Shift, Mapping the Changing Contours of the World Economy.* London: Sage, 2015.

Dickenson, J., 'The naturalist on the River Amazons and a wider world: reflections on the centenary of Henry Walter Bates'. *The Geographical Journal* 158, no. 2(1992): 207–214.

Di Paola, P., *The Knights Errant of Anarchy: London and the Italian Anarchist Diaspora (1880–1917).* Liverpool: Liverpool University Press, 2013.

Di Paola, P., 'Marie Louise Berneri e il gruppo di Freedom Press'. In C. De Maria (ed.), *Maria Luisa Berneri e l'anarchismo inglese,* 133–157. Reggio Emilia: Biblioteca Panizzi/Archivio Famiglia Berneri, 2013.

Dirlik, A., *Anarchism in the Chinese Revolution.* Berkeley: University of California Press, 1991.

Dirlik, A., 'Anarchism and the question of place: thoughts from the Chinese experience'. In S. Hirsch and L. Van der Walt (eds), *Anarchism and Syndicalism in the Colonial and Postcolonial World, 1870–1940,* 131–146. Leiden/Boston: Brill, 2010.

Driver, F., *Geography Militant, Cultures of Exploration and Empire.* Oxford: Blackwell, 2001.

Driver, F., 'In search of the Imperial Map: Walter Crane and the image of Empire'. *History Workshop Journal,* no. 69(2010), 146–157.

Dryhurst, N.F., *Nationalities and Subject Races,* Westminster: King & Son, 1911.

Dugatkin, L., *The Prince of Evolution: Peter Kropotkin's Adventures in Science and Politics.* Charleston: Createspace, 2011.

Dunbar, G. S., 'Some early occurrences of the term "social geography"', *Scottish Geographical Magazine,* no. 93(1977): 15–20.

Dunbar, G.S., *Élisée Reclus: Historian of Nature.* Hamden: Archon Books, 1978.

Dunbar, G.S., *The History of Geography, Collected Essays.* New York: Dodge, 1996.

Dunbar, G.S., 'Élisée Reclus in Louisiana'. In G.S. Dunbar (ed.), *The History of Geography,* 122–131. New York: Dodge, 1996.

Dunbar, G.S., 'Rebecca West and the Reclus Brothers'. In G.S. Dunbar (ed.), *The History of Geography,* 204–213. New York: Dodge, 1996.

Dussel, E., 'Europe, Modernity and Euro-centrism'. *Nepantla. Views from South* 1, no. 3(2000): 465–478.

Ellis, H., 'Élie Reclus'. In J. Ishill (ed.), *The Brothers Élie and Élisée Reclus,* 47–54. Berkeley Heights: Oriole Press, 1927.

Ellis, H., *My Life.* London and Toronto: Heinemann, 1940.

Ellison, J., 'Banging on the walls of Fortress Europe: Tactical media, anarchist politics and border thinking'. In R.J. White, M. Lopes de Souza and S. Springer (eds), *The Practice of Freedom: Anarchism, Geography, and the Spirit of Revolt,* 209–234. London: Rowan and Littlefield, 2016.

Errani, P.L., *Élisée Reclus, geografia sociale.* Milano: Angeli, 1984.

Fabietti, U., *Alle origini dell'antropologia, Tylor, Maine, McLennan, Lubbock, Morgan*. Torino: Boringhieri, 1980.

Farinelli, F. 'Come Lucien Febvre inventó il determinismo'. In L. Febvre, *La Terra e l'Evoluzione Umana*, xi–xxxvii. Turin: Einaudi, 1980.

Farinelli, F., *I segni del mondo*. Firenze: La Nuova Italia, 1992.

Featherstone, D., *Solidarity: Hidden Histories and Geographies of Internationalism*. London: Zed, 2012.

Featherstone, D., 'Black Internationalism, subaltern cosmopolitanism, and the spatial politics of antifascism'. *Annals of the Association of American Geographers* 103 (2013): 1406–1420.

Ferguson, K.E., 'Anarchist women and the politics of walking'. *Political Research Quarterly* 70, no. 4(2017): 708–719.

Ferrari, G.A., *Primo Maggio; storia di un giorno di lotta internazionale sovversivo scomunicato*. Milan: Zero in Condotta, 1986.

Ferretti, F., 'Traduire Reclus: l'Italie écrite par Attilio Brunialti'. *Cybergeo, European Journal of Geography* 14(2009). http://www.cybergeo.eu/index22544.html

Ferretti, F., 'Les Reclus et la Maison Hachette: la première agence de la géographie française?'. *L'Espace Géographique* 39, no. 3(2010): 239–252.

Ferretti, F., 'Comment Élisée Reclus est devenu athée: un nouveau document biographique'. *Cybergeo, revue européenne de géographie* 15(2010). http://cybergeo. revues.org/index22981.html

Ferretti, F., *L'occidente di Élisée Reclus, l'invenzione dell'Europa nella Nouvelle Géographie Universelle (1876–1894)*. Universities of Bologna and Paris 1 Panthéon-Sorbonne: PhD dissertation, 2011.

Ferretti, F., *Élisée Reclus: Lettres de prison et d'exil*. Lardy: Éditions Alafrontière, 2012.

Ferretti, F., 'The correspondence between Élisée Reclus and Pëtr Kropotkin as a source for the history of geography'. *Journal of Historical Geography*, no. 37 (2011): 216–222. doi:10.1016/j.jhg.2010.10.001

Ferretti, F., 'De l'empathie en géographie: la Chine vue par Léon Metchnikoff, Élisée Reclus et François Turrettini'. *Cybergeo European Journal of Geography* (2013). http://cybergeo.revues.org/26127

Ferretti, F., 'They have the right to throw us out: the Élisée Reclus' *Universal Geography*'. *Antipode* 45, no. 5(2013): 1337–1355. doi:10.1111/anti.12006

Ferretti, F., 'Pioneers in the History of Cartography: the Geneva map collection of Élisée Reclus and Charles Perron'. *Journal of Historical Geography*, no. 43(2014): 85–95. doi:10.1016/j.jhg.2013.10.025

Ferretti, F., *Élisée Reclus: pour une géographie nouvelle*. Paris: Éditions du CTHS, 2014.

Ferretti, F., 'Anarchism, geo-history and the origins of the Annales: rethinking Élisée Reclus's influence on Lucien Febvre'. *Environment and Planning D, Society and Space* 33, no. 2(2015): 347–365. doi:10.1068/d14054p

Ferretti, F., 'Radicalizing pedagogy: Geography and libertarian education between the 19th and the 20th century'. In S. Springer, R. White and M. Lopes de Souza, *The Radicalisation of Pedagogy, Anarchism, Geography and the Spirit of Revolt*, 51–72. New York: Rowman & Littlefield, 2016.

Ferretti, F., 'Arcangelo Ghisleri and the "right to barbarity": geography and anti-colonialism in Italy in the Age of Empire (1875–1914)'. *Antipode* 48, no. 3(2016): 563–583. doi:10.1111/anti.12206

Ferretti, F., 'Organisation and formal activism: insights from the anarchist tradition'. *International Journal of Sociology and Social Policy* 36, no. 11–12(2016): 726–740. doi:10.1108/ijssp-11-2015-0127

Ferretti, F., 'Anarchist geographers and feminism in late 19th century France: the contributions of Élisée and Élie Reclus'. *Historical Geography* 44(2016): 68–88.

Ferretti, F., 'Evolution and revolution: anarchist geographies, modernity and post-structuralism'. *Environment and Planning D-Society and Space* 35, no. 5(2017): 893–912. doi:10.1177/0263775817694032

Ferretti, F., 'Publishing anarchism: Peter Kropotkin and British print cultures, 1876–1917'. *Journal of Historical Geography*, no. 57(2017): 17–27. doi:10.1016/j.jhg.2017.04.006

Ferretti, F., 'Political geographies, unfaithful translations and anti-colonialism: Ireland in Élisée Reclus's geography and biography'. *Political Geography*, no. 59 (2017): 11–23.

Ferretti, F., 'Tropicality, the unruly Atlantic and social utopias: the French explorer Henri Coudreau (1859–1899)'. *Singapore Journal of Tropical Geography* 38, no. 3 (2017): 332–349. doi:10.1111/sjtg.12209

Ferretti, F., 'The murderous civilisation: anarchist geographies, ethnography and cultural differences in the works of Élie Reclus'. *Cultural Geographies* 24, no. 1(2017): 111–129.

Ferretti, F., 'Revolutions and their places', the anarchist geographers and the problem of nationalities in the Age of Empire (1875–1914)' In F. Ferretti, G. Barrera de la Torre, A. Ince and F. Toro (eds), *Historical Geographies of Anarchism: Early Critical Geographers and Present-Day Scientific Challenges*, 113–128. London: Routledge, 2017.

Ferretti, F. 'Situated knowledge and visual education: Patrick Geddes and Reclus's geography (1886–1932)'. *Journal of Geography* 116, no. 1(2017): 3–19.

Ferretti, F., 'Teaching anarchist geographies: Élisée Reclus in Brussels and 'the art of not being governed'. *Annals of the American Association of Geographers* 108, no. 1 (2018): 162–178.

Ferretti, F., 'Geographies of internationalism: radical development and critical geopolitics from the Northeast of Brazil'. *Political Geography*, no. 63(2018): 10–19.

Ferretti, F. and P. Minder, *«Pas de la dynamite, mais du tabac». L'enquête de 1885 contre les anarchistes en Suisse romande.* Paris: Éditions du Monde Libertaire, 2015.

Ferretti, F., P. Malburet and P. Pelletier, *Élisée Reclus et les Juifs, un géographe anarchiste face à une question brûlante.* Paris: L'Harmattan, 2018.

Fichman, M., *An Elusive Victorian: The Evolution of Alfred Russel Wallace.* Chicago and London: The University of Chicago Press, 2004.

Fleming, M., *The Geography of Freedom. The Odyssey of Élisée Reclus.* Montréal and New York: Black Rose Books, 1988.

Floud, R.C., 'Britain 1860–1914, a survey'. In R.C. Floud and N. McCloskey (eds), *The Economic History of Britain since 1700, vol. 2, 1860 to the 1970s*, 1–26. Cambridge: Cambridge University Press, 1981.

Fowle, F. and Thomson, B. (Eds), *Patrick Geddes: the French Connection.* Oxford: White Cockade, 2004.

Fox Bourne, H.R., 'The claims of uncivilised races'. *The Humane Review*, no. 1(1900): 162–172.

Freedom Press, *Our First Centenary.* London: Freedom Press, 1986. https://libcom.org/files/freedom-centenary.pdf

Fyfe, A., 'Journals, learned societies and money, Philosophical Transactions, ca. 1750–1900'. *Notes and Records of the Royal Society* 69(2015): 277–299.

Geddes, P., *Cities in Evolution*. London: Williamson & Norgate, 1949.

Geddes, P., J.B. Jordan and H.F. Brin, 'A great globe: Discussion'. *The Geographical Journal*, no 12(1898): 406–409. doi:10.2307/1774766

Girón Sierra, A., 'Kropotkin between Lamarck and Darwin: the impossible synthesis'. *Asclepio*, no. 55(2003): 189–213. doi:10.3989/asclepio.2003.v55.i1.94

Goldman, E., *Living my Life*. New York: Knopf, 1931.

Goldman, E., 'The tragedy of the political exiles'. *The Nation*, 10 October, 1934. https://theanarchistlibrary.org/library/emma-goldman-the-tragedy-of-the-political-exiles

Gomme, R., *George Herbert Perris 1866–1920, the Life and Times of a Radical*. Oxford: Peter Lang, 2003.

Goodway, D., *Anarchist Seeds Beneath the Snow. Left-Libertarian Thought and British Writers from William Morris to Colin Ward*. Liverpool: Liverpool University Press, 2006.

Gould, S.J. 'Kropotkin was no crackpot'. *Natural History* 106(1997): 12–21.

Graeber, D., *Fragments of an Anarchist Anthropology*. Chicago: Prickly Paradigm Press, 2004.

Graham, R. *We Do Not Fear Anarchy, We Invoke It: The First International and the Origins of the Anarchist Movement*. London: AK Press, 2015.

Grave, J., (ed.) *Patriotisme et colonisation*. Paris: Les editions du Temps Nouveaux, 1903.

Grave, J., *Mémoires d'un anarchiste*. Paris: Editions du Sextant, 2009.

Gregory, D., *The Colonial Present: Afghanistan, Palestine, Iraq*. Oxford: Blackwell, 2004.

Gregory, J., *Victorians Against the Gallows, Capital Punishment and the Abolition Movement in Nineteenth-Century Britain*. London: Tauris, 2012.

Griffin, P., 'Making usable pasts: collaboration, labour and activism in the archive'. *Area*, 2017 [early view]. doi:10.1111/area.12384

Gwynn, R., *The Huguenots of London*. Brighton: Alpha Press, 1998.

Haaland, B., *Emma Goldman, Sexuality and the Impurity of the State*. Montreal: Black Rose Books, 1993.

Hale, P.J. *Political Descent: Malthus, Malthusianism and the Politics of Evolution in Victorian England*. Chicago and London: Chicago University Press, 2014.

Haraway, D., *Simians, Cyborgs, and Women. The Reinvention of Nature*. New York: Routledge, 1991.

Harris, C., 'Archival fieldwork'. *Geographical Review*, no. 91(2001): 328–334. doi:10.2307/3250834

Harris, M., *Rebellion on the Amazon: the Cabanagem, Race, and Popular Culture in the North of Brazil, 1798–1840*. New York: Cambridge University Press, 2010.

Harvey, D., '"Listen, Anarchist! A personal response to Simon Springer's "Why a radical geography must be anarchist"'. *Dialogues in Human Geography* 7, no. 3 (2017): 233–250. doi:10.1177/2043820617732876

Heath, C. 'Francisco Ferrer'. *The Humane Review*, no. 10(1909): 193–197.

Heath, N., 'Alfred Marsh 1858–1914'. *Libcom*, 2011. http://ns210054.ovh.net/history/marsh-alfred-1858-1914

Heath, N., 'Agnes Henry 1850–1915'. *Libcom*, 2011. https://libcom.org/history/henry-agnes-1850-1915

Heath, N., 'Nannie Florence Dryhurst 1856–1930'. *Libcom*, 2012. https://libcom.org/history/dryhurst-nannie-florence-1856-1930

Heath, R., *The English Peasant*. London: Fisher & Unwin, 1892.

Heath, R., 'The Kafir and his master'. *The Humane Review* 1(1900): 72–76.

Heath, R., 'Élisée Reclus'. *The Humane Review* 6(1905): 129–142.

Hecht, J.M., *The End of the Soul, Scientific Modernity, Atheism, and Anthropology in France*. New York: Columbia University Press, 2003.

Hecht, S.B., *The Scramble for the Amazon and the 'Lost Paradise' of Euclides da Cunha*. Chicago: The University of Chicago Press, 2013.

Hechter, M., *Internal Colonialism: The Celtic Fringe in British National Development, 1536–1966*. London: Routledge and Kegan Paul, 1978.

Henderson, P., *William Morris, his Life Works and Friends*. London/New York: Longmans, 1967.

Hendrick, G., *Henry Salt, Humanitarian Reformer and Man of Letters*. Urbana/London: University of Illinois Press, 1977.

Hinely, S., 'Charlotte Wilson, the "Woman Question", and the meanings of anarchist socialism in late Victorian radicalism'. *IRSH*, no. 57(2012): 3–36. doi:10.1017/S0020859011000757

Hirsch, S. and L. Van der Walt (eds), *Anarchism and Syndicalism in the Colonial and Postcolonial World, 1870–1940: The Praxis of National Liberation, Internationalism, and Social Revolution*, Leiden/Boston: Brill, 2007.

Hodder, J., S. Legg and M. Heffernan, 'Introduction: historical geographies of internationalism, 1900–1950'. *Political Geography*, no. 49(2015): 1–6.

Homobono, J.I., 'Las ciudades y su evolución: análisis del fenómeno urbano en la obra de Élisée Reclus'. *Zainak*, no. 31(2009): 75–116

Hug, H., *Peter Kropotkin (1842–1921), Bibliographie*. Berlin: Edition Anares im Trotzdem-Verlag, 1994.

Hugo, V., *Les misérables*. Paris: Lacroix, 1862.

Hulse, J.W., *Revolutionists in London. A Study of Five Unorthodox Socialists*. Oxford: Clarendon Press.

Hyndman, H., *The Record of an Adventurous Life*. London: MacMillan, 1911.

Ishill, J. (ed.), *Peter Kropotkin: The Rebel Thinker and Humanitarian*, Berkeley Heights: Oriole Press, 1923.

Ishill, J. (ed.), *Élisée and Élie Reclus: in Memoriam*. Berkeley Heights: Oriole Press, 1927.

John, A., *War, Journalism and the Shaping of the Twentieth Century: The Life and Times of Henry W. Nevinson*. London/New York: I.B. Tauris, 2006.

Jordan, J., *Josephine Butler*. London: Murray, 2001.

Jordan, J. and I. Sharp, *Josephine Butler and the Prostitution Campaigns Diseases of the Body Politic*. London: Routledge, 2004.

Jöns, H., P. Meusburger and M. Heffernan, *Mobilities of Knowledge*. Bern: Springer, 2017.

Kaplan, R., *The Revenge of Geography*. New York: Random House, 2012.

Kearns, G., 'The political pivot of geography'. *The Geographical Journal*, no. 170 (2004): 337–346. doi:10.1111/j.0016-7398.2004.00135.x

Kearns, G., *Geopolitics and Empire: The Legacy of Halford Mackinder*. Oxford: Oxford University Press, 2009.

Kearns, G., Meredith, D. and J. Morrissey (eds), *Spatial Justice and the Irish Crisis*. Dublin: Royal Irish Academy, 2014

Keighren, I.M., *Bringing Geography to Book: Ellen Semple and the Reception of Geographical Knowledge*. London: IB Tauris, 2010.

Keighren, I.M., 'History and philosophy of geography I: The slow, the turbulent, and the dissenting'. *Progress in Human Geography* 41, no. 5(2016), 638–647.

Keighren, I.M., 'History and philosophy of geography II: The excluded, the evil, and the anarchic'. *Progress in Human Geography* (2017) [early view]. doi:10.1177/0309132517730939

Keighren, I.M., C.W.J. Withers and B. Bell, *Travels into Print: Exploration, Writing, and Publishing with John Murray, 1773–1859*. Chicago: Chicago University Press, 2015.

Kinna, R., *William Morris: The Art of Socialism*. Cardiff: University of Wales Press, 2000.

Kinna, R., *Kropotkin: Reviewing the Classical Anarchist Tradition*. Edinburgh: Edinburgh University Press, 2016.

Kissack, T., *Free Comrades. Anarchism and Homosexuality in the United States, 1885–1917*. Oakland: AK Press, 2008.

Konishi, S., *Anarchist Modernity: Cooperatism and Japanese-Russian Intellectual Relations in Modern Japan*. Cambridge: Harvard University Press, 2013.

Kropotkin, P., 'What geography ought to be'. *The Nineteenth Century*, no. 18(1885): 940–956.

Kropotkin, P., *In Russian and French Prisons*. London: Ward and Doney, 1887.

Kropotkin, P., 'Tourgueneff, Tolstoi, and Dostoievsky', *The Scottish Art Review*, no. 2 (1889): 150–153.

Kropotkin, P., 'Natural selection and mutual aid'. In E. Carpenter (ed.), *Humane Science Lectures*, 182–186. London: George Bell & Sons, 1897.

Kropotkin, P., *Anarchism, its Philosophy and Ideal*, San Francisco: Free Press, 1898.

Kropotkin, P., *Fields, Factories and Workshops*. London: Hutchinson, 1898.

Kropotkin, P., *Memoirs of a Revolutionist, vol. II*. London: Smith, Elder & Co., 1899.

Kropotkin, P., *Mutual aid, a factor in evolution*. London: Heinemann, 1902.

Kropotkin, P., *Modern Science and Anarchism*. Philadelphia: The Social Science Club, 1903.

Kropotkin, P., 'The Orography of Asia I. Introductory remarks'. *The Geographical Journal*, no. 2(1904): 176–204.

Kropotkin, P., 'The Orography of Asia'. *The Geographical Journal*, no. 3(1904): 331–361.

Kropotkin, P., *Russian Literature: Ideals and Realities*. London: Duckworth, 1905.

Kropotkin, P., *The Great French Revolution: 1789–1793*. London: Heinemann, 1909.

Kropotkin, P., *Ethics, Origin and Development*. London: Harrap, 1924.

Kropotkin, P., *Act for Yourselves, Articles from Freedom 1886–1907*. London: Freedom Press, 1998.

Lacoste, Y., 'Élisée Reclus. Une très large conception de la géographicité et une bienveillante géopolitique'. *Hérodote*, no. 117(2005), 29–52.

Landuyt, I. and G. Lernout, 'Joyce's sources: Les Grands Fleuves Historiques'. *Joyce Studies Annual* 6(1995), 99–138.

Lankester, E.R., 'Heredity and the direct action of the environment'. *The Nineteenth Century and After*, 68, no. 403(1910): 483–491.

Latour, B., *Science in Action: How to Follow Scientists and Engineers through Society*. Cambridge: Harvard University Press, 1987.

Laursen, B.O., 'Anarchist anti-imperialism: Guy Aldred and the Indian Revolutionary Movement, 1909–1914'. *Journal of Imperial and Commonwealth History* (2018).

La Vergata, A., *Colpa di Darwin ? Razzismo, eugenetica, guerra e altri mali*. Turin: UTET, 2009.

Levy, C., 'Foreword'. In V. Richards, *The Anarchist Writings of Errico Malatesta*. Oakland: PM Press, 2015.

Ley, D., *Geography Without Man: A Humanistic Critique*. University of Oxford: Research Paper, no. 24, 1980.

Livingstone, D.N., *The Geographical Tradition*. Oxford and Cambridge: Wiley, 1992.

Livingstone, D.N., *Putting Science in its Place: Geographies of Scientific Knowledge*. Chicago: Chicago University Press, 2003.

Livingstone, D.N., 'Science, text and space: thoughts on the geography of reading'. *Transactions of the Institute of British Geographers* 30, no. 4(2005): 391–401. doi:10.1111/j.1475-5661.2005.00179.x

Livingstone, D.N., 'Changing climate, human evolution, and the revival of environmental determinism'. *Bulletin of History of Medicine* 86, no. 4(2012):564–595. doi:10.1353/bhm.2012.0071

Livingstone, D.N., *Dealing with Darwin*. Baltimore: Johns Hopkins University Press, 2014.

Livingstone, D.N., 'Finding revelation in anthropology: Alexander Winchell, William Robertson Smith and the heretical imperative'. *The British Journal for the History of Science*, no. 48(2015): 435–454. doi:10.1017/s0007087415000035

Livingstone, D.N., 'Debating Darwin at the Cape'. *Journal of Historical Geography*, no. 52(2016): 1–15. doi:10.1016/j.jhg.2015.12.002

Lomnitz, C., *The Return of Comrade Ricardo Flores Magón*. New York: Zone Books, 2014.

Lorimer, H. and C. Philo, 'Disorderly archives and orderly accounts: reflections on the occasion of Glasgow's Geographical Centenary'. *Scottish Geographical Journal* 125, no. 3–4(2009): 227–255.

MacCarthy, F., *William Morris: A Life for our Time*. London: Faber & Faber, 1994.

MacCarthy, F., *Anarchy and Beauty, William Morris and his Legacy*. London: National Portrait Gallery, 2014.

Mächler-Tobar, E., *Un nombre expoliado: Élisée Reclus y su visión de América*. Bogotá: Editorial Universidad del Rosario, 2014.

MacLaughlin, J., *Kropotkin and the Anarchist Intellectual Tradition*. London: Pluto Press, 2016.

Malatesta, E., 'The most greatly humane man'. In J. Ishill (ed.), *Peter Kropotkin: The Rebel Thinker and Humanitarian*, 38–39. Berkeley Heights: Oriole Press, 1923.

Manfredonia, G., *Les anarchistes et la Révolution Française*. Paris: Les Editions du Monde Libertaire, 1990.

Manton, K., 'The Fellowship of the New Life: English Ethical Socialism reconsidered'. *History of Political Thought* 24, no. 2(2003): 282–304.

Marchant, J., *Alfred Russel Wallace: Letters and Reminiscences*. London: Cassell, 1916.

Marshall, P., *Demanding the Impossible: A History of Anarchism*. London: Harper Collins, 1992.

Massey, D., *World City*. Cambridge: Polity, 2007.

Mavor, J., *My Windows on the Streets of the World*. London and Toronto: Dent, 1923.

Maxwell, B. and R. Craib (eds), *No Gods no Masters no Peripheries, Global Anarchisms*. Oakland: PM Press, 2015.

Mayhew, R., 'Materialist hermeneutics, textuality and the history of geography: print spaces in British geography'. *Journal of Historical Geography* 33(2007): 466–488.

Mayhew, R.J., *Malthus: The Life and Legacies of an Untimely Prophet*. Cambridge: Harvard University Press, 2014.

McCoole, S., *Easter Widows: Seven Irish Women Who Lived in the Shadow of the 1916 Rising*. Dublin: Doubleday Ireland, 2014.

McCormack, W.J., *Roger Casement in Death or Haunting the Free State*. Dublin: UCD Press, 2002.

McGregor, J. 'Locating exile: decolonization, anti-imperial spaces and Zimbabwean students in Britain, 1965–1980'. *Journal of Historical Geography* 57(2017): 62–75.

McKay, I., 'Kropotkin, Woodcock and Les Temps Nouveaux'. *Anarchist Studies* 23, no. 1(2015): 62–85.

Mečnikov, L.I., 'Revolution and evolution'. *Contemporary Review*, no. 50(1886): 412–437.

Meller, H., *Patrick Geddes: Social Evolutionist and City Planner*. London: Routledge, 1990.

Metcalf, P., *James Knowles Victorian Editor and Architect*. Oxford: Clarendon Press, 1980.

Michel, L., *Exile en Nouvelle Calédonie*. Paris: Magellan, 2005.

Mill, H.R. and D. Freshfield, 'Obituary: Sir John Scott Keltie'. *The Geographical Journal*, no. 69(1927): 281–286.

Mill, H.R., *An Autobiography*. London: Longmans, 1951.

Miller, M., *Kropotkin*. Chicago: Chicago University Press, 1976.

Minca, C., 'Humboldt's compromise, or the forgotten geographies of landscape'. *Progress in Human Geography*, no. 31(2007): 179–193.

Morris, B., *The Anarchist Geographer: An Introduction to the Life of Peter Kropotkin*. Minehead: Genge Press, 2007.

Morris, W., *The Collected Letters of William Morris, Vol. 2, Part b, 1885–1888*. Princeton: Princeton University Press, 1987.

Morris, W., *The Collected Letters of William Morris, Vol. 3, 1889–1892*. Princeton: Princeton University Press, 1996.

Morris, W., *The Collected Letters of William Morris, Vol. 4, 1893–1896*. Princeton: Princeton University Press, 1996.

Mumford, L., *An Introduction to the History of Sociology*. Chicago: Chicago University Press, 1966.

Muthu, S., *Enlightenment Against Empire*. Princeton: Princeton University Press, 2003.

Naylor, S., 'Historical geography: knowledge, in place and on the move'. *Progress in Human Geography*, no. 29(2005): 626–634. doi:10.1191/0309132505ph573pr

Nettlau, M., 'Peter Kropotkin at work'. In J. Ishill (ed.), *Peter Kropotkin: the Rebel Thinker and Humanitarian*, 11–18. Berkeley Heights: Oriole Press, 1923.

Nettlau, M., *Élisée Reclus, vida de un sabio justo y rebelde, vol. 1*. Barcelona: Ediciones de la Revista Blanca, 1928.

Nettlau, M., *Élisée Reclus, vida de un sabio justo y rebelde, vol. 2*. Barcelona: Ediciones de la Revista Blanca, 1930.

Nevin, D., *James Connolly, a Full Life*. Dublin: Gill & Macmillan, 2005.

Nevinson, H., *Fire of Life*. London: James Nisbet, 1935.

Newsinger, J., 'Why Rhodes must fall'. *Race & Class* 58, no. 2(2016): 70–78. doi:10.1177/0306396816657726

Notehelfer, F.G., *Kōtoku Shusui: Portrait of a Japanese Radical*. Cambridge: Cambridge University Press, 1971.

Nozawa, H., 'Development of Ishikawa Sanshiro's anarchism under the influence of Élisée Reclus'. *Geographical Review of Japan* 79, no. 14(2006): 837–856.

Ogborn, M., *Indian Ink: Script and Print in the Making of the English East India Company*. Chicago: Chicago University Press, 2007.

Ogborn, M. and C.W.J. Withers (eds), *Geographies of the Book*. Farnham: Ashgate, 2010.

Oyón, J.L. and M. Serra, 'Las casas de Reclus: hacia la fusión naturaleza-ciudad, 1830–1871'. *Scripta Nova*, no. 421(2012) http://www.ub.edu/geocrit/sn/sn-421.htm

Owen, J., *Darwin's Apprentice: An Archaeological Biography of John Lubbock*. Barnsley: Pen & Sword Books, 2013.

Papillard, F., *Michelet et Vascœuil*. Paris: Amis du Château de Vascœuil et de Michelet, 1974.

Pearson, A. and M. Heffernan, 'The American Geographical Society's map of Hispanic America: million-scale mapping between the wars'. *Imago Mundi*, no 61 (2009): 215–243.

Pease, M.R., *Richard Heath. 1831–1912*. Letchworth: Garden City Press, 1922.

Pedgley, D.E., 'Mill, Hugh Robert (1861–1950)'. In *Oxford Dictionary of National Biography*, 2004, https://doi.org/10.1093/ref:odnb/35021

Pelletier, P., *Géographie et anarchie: Reclus, Kropotkine, Metchnikoff*. Paris: Editions du Monde Libertaire, 2013.

Pelletier, P., *Kôtoku Shûsui: socialiste et anarchiste japonais*. Paris: Editions du Monde Libertaire, 2015.

Pesce, G., *Da ieri a domani. La pianificazione organica di Kropotkin, Reclus, Branford e Geddes, Mumford*. Bologna: Clueb, 1980.

Preece, R., *Animal Sensibility and Inclusive Justice in the Age of Bernard Shaw*. Vancouver: UBC Press, 2011.

Pramopolini, C. *Predica di Natale: opuscolo di propaganda per le campagne*. Reggio Emilia: Tipografia Operaia, 1899.

Prichard, A., *Justice, Order and Anarchy: The International Political Theory of Pierre-Joseph Proudhon*. London: Routledge, 2013.

Purchase, G., *Peter Kropotkin, Ecologist, Philosopher and Revolutionary*. Sydney, University of South Wales: PhD Dissertation, 2003.

Quail, J., *The Slow Burning Fuse: The Lost History of the British Anarchists*. London: Paladin, 1978.

Raghuram, P. and C. Madge, 'Towards a method for postcolonial development geography? Possibilities and challenges'. *Singapore Journal of Tropical Geography*, no. 27(2006): 270–288.

Ramnath, M., *Decolonizing Anarchism: An Antiauthoritarian History of India's Liberation Struggle*. London: AK Press, 2011.

Reclus, Élie, 'Female kinship and maternal filiation'. *Radical Review* 1, no. 11(1877): 205–223.

Reclus, Élie, 'Ethnography and ethnology'. In *Encyclopaedia Britannica, vol. 8*, 613–626. Edinburgh, 1878.

Reclus, Élie, 'The evil eye'. *Cornhill Magazine*, no. 39(1879): 184–198.

Reclus, Élie, 'The fire'. In *Encyclopaedia Britannica, vol. 9*, 227–232. Edinburgh, 1879.

Reclus, Élie, *Les primitifs*. Paris: Chamerot, 1885.

Reclus, Élisée, 'Introduction'. In C. Ritter, 'De la configuration des continents sur la surface du globe et de leurs fonctions dans l'histoire'. *Revue Germanique*, no. 11 (1859), 259.

Reclus, Élisée, *Guide du voyageur à Londres et aux environs*. Paris: Hachette: 1860.

Reclus, Élisée, *Londres illustré, guide spécial pour l'exposition de 1862*. Paris: Hachette, 1862.

Reclus, Élisée, *Du sentiment de la nature dans les sociétés modernes*. Paris: Extrait de la *Revue des Deux Mondes*, 1866.

Reclus, Élisée, *The Earth: A Descriptive History of the Phenomena of the Life of the Globe*. London: Chapman and Hall, 1871.

Reclus, Élisée, 'Hégémonie de l'Europe'. *La Société Nouvelle*, no. 112(1894): 433–443.

Reclus, Élisée, 'Projet de construction d'un globe terrestre à l'échelle du 100.000e'. In *Report of the Sixth International Geographical Congress held in London in 1896*, 625–636. London: John Murray, 1895.

Reclus, Élisée, 'A great globe'. *The Geographical Journal*, no. 12(1898): 401–406.

Reclus, Élisée, 'Léopold de Saussurre, Psychologie de la colonisation française dans ses rapports avec les sociétés indigènes, Paris, Alcan, 1899'. *L'Humanité Nouvelle*, 26 (1899): 246–248.

Reclus, Élisée, 'On spherical maps and reliefs'. *The Geographical Journal*, no. 3(1903): 290–293.

Reclus, Élisée, Mr. Mackinder, Mr. Ravenstein, Dr. Herbertson, Prince Kropotkin, Mr. Andrews, Cobden Sanderson, 'On spherical maps and reliefs. Discussion'. *The Geographical Journal*, no. 3(1903): 294–299.

Reclus, Élisée, 'The great kinship'. *The Humane Review* 6(1905): 206–214.

Reclus, Élisée, *L'Homme et la Terre, vol. V*. Paris: Librairie Universelle, 1905.

Reclus, Élisée, *L'Homme et la Terre, vol. VI*. Paris: Librairie Universelle, 1908.

Reclus, Élisée, *Correspondance, vol. I*. Paris: Schleicher, 1911.

Reclus, Élisée, *Correspondance, vol. II*. Paris: Schleicher, 1911.

Reclus, Élisée, *Correspondance, vol. III*. Paris: Costes, 1925.

Reclus, Élisée, 'Élie Reclus'. In J. Ishill (ed.), *The Brothers Élie and Élisée Reclus*, 241–271. Berkeley Heights: Oriole Press, 1927.

Regard, F. (ed.), *Féminisme et prostitution dans l'Angleterre du XIXe siècle: la croisade de Josephine Butler*. Lyons: ENS Editions, 2013.

Reynaud-Paligot, C., *La république raciale: paradigme racial et idéologie républicaine, 1860–1930*. Paris: PUF, 2006.

Romani, C., *Oreste Ristori: uma aventura anarquista*. São Paulo: Annablume, 2002.

Rowbotham, S. and J. Weeks, *Socialism and the New Life: The Personal and Sexual Politics of Edward Carpenter and Havelock Ellis*. London: Pluto Press, 1977.

Russell, B., *Roads to Freedom: Socialism, Anarchism, and Syndicalism*. London: Allen and Unwin, 1918.

Sacchetti, G., *Eretiche, il Novecento di Maria Luisa Berneri e Giovanna Caleffi*. Milano: Biblion, 2017.

Said, E., *Orientalism*. New York: Vintages Books, 1978.

Said, E., 'Afterword: reflections on Ireland and Post-colonialism'. In C. Carroll and P. King (eds), *Ireland and Postcolonial Theory*, 177–185. Cork: Cork University Press, 2003.

Salt, H.S., 'Introductory'. *The Humane Review*, no. 1(1900): 1–2.

Salt, H.S., 'Access to mountains'. *The Humane Review*, no. 9(1908): 247–252.

Salt, H.S., 'Concerning cannibalism'. *The Humane Review*, no. 10(1909): 247–252.

Salt, H.S., 'A poet of socialism'. *The Humane Review* (April1910): 13–23.

Salt, H.S., *Seventy Years among Savages*. London: George Allen & Unwin, 1921.

Salt, H.S., 'The many-sided man of genius'. In J. Ishill (ed.), *The Brothers Élie and Élisée Reclus*, 67–68. Berkeley Heights: Oriole Press, 1927.

Sellers, E., 'Our most distinguished refugee'. *The Contemporary Review* 66(1894): 537–549.

Scott, J.C., *The Art of Not Being Governed: An Anarchist History of Upland Southeast Asia*. New Haven: Yale University Press, 2009.

Scott Keltie, J., *Geographical Education. Report to the Council of the Royal Geographical Society*. London, 1885.

Scott Keltie, J., 'Obituary: Prince Kropotkin'. *The Geographical Journal*, no. 57(1921): 316–319.

Secord, J., *Victorian Sensation: The Extraordinary Publication, Reception, and Secret Authorship of Vestiges of the Natural History of Creation*. Chicago: Chicago University Press, 2000.

Secord, J., 'Knowledge in transit'. *Isis*, no. 95(2004), 654–672. doi:10.1086/430657

Secord, J., *Visions of Science*. Oxford: Oxford University Press, 2014.

Shaffer, K., 'Tropical libertarians: anarchist movements and networks in the Caribbean, Southern United States, and Mexico, 1890s–1920s'. In S. Hirsch and L. Van der Walt (eds), *Anarchism and Syndicalism in the Colonial and Postcolonial World, 1870–1940*, 273–320. Leiden/Boston: Brill, 2010.

Shaffer, K., 'Latin lines and dots: transnational anarchism, regional networks, and italian libertarians in Latin America'. *Zapruder World*, no. 1(2014). http://zapruder world.org/journal/archive/volume-1/latin-lines-and-dots-transnational-anarchism-regional-networks-and-italian-libertarians-in-latin-america/

Sidaway, J.D., Mamadouh, V. and M. Power. 'Reappraising geopolitical traditions'. In K. Dodds, M. Kuus and J. Sharp (eds), *The Ashgate Research Companion to Critical Geopolitics*, 165–187. Farnham: Ashgate, 2013.

Siegrist, P., 'Historicising anarchist geographies: six issues for debate from a historian's point of view'. In F. Ferretti, G. Barrera, A. Ince, F. Toro (eds), *Historical Geographies of Anarchism*, 129–150. London: Routledge, 2017.

Smith, N., *Uneven Development: Nature, Capital and the Production of Space*. Oxford: Blackwell, 1984.

Souza, M.L. de, 'Lessons from praxis: autonomy and spatiality in contemporary Latin American social movements'. *Antipode* 48, no. 5(2016): 1292–1316. doi:10.1111/ anti.12210

Souza, M.L. de, R.J. White, and S. Springer (eds), *Theories of Resistance: Anarchism, Geography and the Spirit of Revolt*. Lanham: Rowman & Littlefield, 2016.

Springer, S., 'The Limits to Marx. David Harvey and the Condition of Postfraternity'. *Dialogues in Human Geography* 7, no. 3(2017), 280–294. doi:10.1177/ 2043820617732918

Springer, S., *The Anarchist Roots of Geography: Toward Spatial Emancipation*. Minneapolis: University of Minnesota Press, 2016.

Springer, S., 'Earth Writing'. *GeoHumanities* 3, no. 1(2017): 1–19. doi:10.1080/ 2373566X.2016.1272431

Springer, S., A.J. Barker, G. Brown, A. Ince, and J. Pickerill, 'Reanimating anarchist geographies: A new burst of colour'. *Antipode*, no. 44(2012): 1591–1604.

Stahler-Sholk, R., H. Vanden, and M. Becker (eds), *Rethinking Latin American Social Movements Radical Action from Below*. London: Rowman & Littlefield, 2014.

Startt, J.D., 'The evolution of an anarchist: an autobiographical statement by Varlaam Tcherkesoff, 1846–1925'. *Biography* 10, no. 2(1987): 142–150. doi:10.1353/ bio.2010.0490

Steele, T. 'Élisée Reclus et Patrick Geddes géographes de l'esprit'. *Réfractions*, no. 4 (1999). https://raforum.info/reclus/spip.php?article26

Stocking, G.W., 'What's in a name? The origins of the Royal Anthropological Institute (1837–71)'. *Man* 6, no. 3(1971): 369–390.

Stocking, G.W., *Victorian Anthropology*. London: Collier MacMillan, 1987.

Stocking, G.W., *After Tylor, British Social Anthropology*. London: Athlone, 1996.

Stoddart, D., *On Geography and its History*. Oxford: Basil Blackwell, 1986.

Sue, E., 'Les mystères de Paris'. *Journal des débats*, 9 June 1842 until 15 October 1843.

Tang, C., *The Geographic Imagination of Modernity*. Stanford: Stanford University Press, 2008.

Tcherkesoff, V., *Pages of Socialist History*. New York: Cooper, 1902.

Tcherkesoff, V., 'Friend and comrade'. In J. Ishill (ed.), *Peter Kropotkin: The Rebel Thinker and Humanitarian*, 23–26. Berkeley Heights: Oriole Press, 1923.

Thompson, E.P., *The Making of the English Working Class*. London: Gollancz, 1963.

Thompson, E.P., *William Morris: Romantic to Revolutionary*. London: Merlin Press, 1977.

Thomson, M.A., 'The humane study of natural history'. In E. Carpenter (ed.), *Humane Science Lectures*, 35–76. London: George Bell & Sons, 1897.

Tierney, R.T., *Monster of the Twentieth Century: Kôtoku Shûsui and Japan's First Anti-Imperialist Movement*. Oakland: University of California Press, 2015.

Tombs, R., *The Paris Commune, 1871*. London/New York: Longman, 1999.

Tsuzuki, C., *Edward Carpenter, 1844–1929: Prophet of Human Fellowship*. London/New York/Melbourne: Cambridge University Press, 1980.

Turcato, D., 'Italian anarchism as a transnational movement, 1885–1915'. *International Review of Social History* 52, no. 3(2007): 407–444. doi:10.1017/s0020859007003057

Turcato, D. (ed.), *An Errico Malatesta Reader*. London: AK Press, 2014.

Turcato, D., *Making Sense of Anarchism: Errico Malatesta's Experiments with Revolution*. London: AK Press, 2015.

Van der Walt, L., 'Revolutionary syndicalism, communism and the national question in South African socialism, 1886–1928'. In S. Hirsch and L. Van der Walt (eds), *Anarchism and Syndicalism in the Colonial and Postcolonial World, 1870–1940*, 33–94. Boston/Leiden: Brill, 2010.

Varengo, S., *Pagine anarchiche: Pëtr Kropotkin e il mensile Freedom (1886–1914)*. Milan: Biblion, 2015

Véron, O., '(Extra)ordinary activism: veganism and the shaping of hemeratopias'. *International Journal of Sociology and Social Policy* 36, no. 11/12(2016): 756–773.

Volin, V., *The Unknown Revolution, 1917–1921*. Detroit: Black & Red, 1974.

Wallace, A.R., 'The proposed gigantic model of the earth'. *Contemporary Review*, no. 69(1896): 73–74.

Wallace, A.R., *Studies. Scientific and social, vol. II*. London: Macmillan, 1900.

Wallace, A.R., *My Life: A Record of Events and Opinions, vol. II*. London: Chapman & Hall, 1905.

Walter, N., 'Guy A. Aldred'. *The Raven*, no. 1(1986): 77–92.

Walter, N., 'Introduction'. In C. Wilson, *Anarchist Essays*. London: Freedom Press, 2000.

Walter, N. and H. Becker, 'Introduction'. In P. Kropotkin, *Act for Yourselves, Articles from Freedom 1889–1907*, 7–18. London: Freedom Press, 1988.

Wang, Y., H. Liu and Z. Sun, 'Lamarck rises from his grave: parental environment-induced epigenetic inheritance in model organisms and humans'. *Biological Review*, no. 92(2017): 2084–2111.

Warketin, J., 'James Mavor, forerunner in Canadian geography'. *The Canadian Geographer* 58, 3(2014): 377–392. doi:10.1111/cag.12082

Weinbren, D., 'Against all cruelty: the Humanitarian League, 1891–1919'. *History Workshop*, no. 38(1994): 86–105.

Welter, V., *Biopolis: Patrick Geddes and the City of Life.* Cambridge: The MIT Press, 2002.

White, R.J., 'Explaining why the non-commodified sphere of mutual aid is so pervasive in the advanced economies'. *International Journal of Sociology and Social Policy* 29, no. 9/10(2009): 457–472. doi:10.1108/01443330910986252

White, R.J., 'Animal geographies, anarchist praxis, and critical animal studies'. In K. Gillespie and R.C. Collard (eds), *Critical Animal Geographies: Politics, Intersections and Hierarchies in a Multispecies World*, 19–35. London/New York: Routledge, 2015.

White, R.J., 'Following in the footsteps of Élisée Reclus: disturbing places of interspecies violence that are hidden in plain sight'. In A.J. Nocella II, R.J. White and E. Cudworth (eds), *Anarchism and Animal Liberation. Essays on Complementary Elements of Total Liberation*, 212–230. Jefferson: Mc Farland & Company, 2015.

White, R.J. and C. Williams, 'The pervasive nature of heterodox economic spaces at a time of neoliberal crisis: towards a "postneoliberal" anarchist future'. *Antipode* 44, no. 5(2012): 1625–1644. doi:10.1111/j.1467-8330.2012.01033.x

Willems, N., 'Contesting imperial geography: reading Élisée Reclus in 1930s' Hokkaido'. In R.J. White, M. Lopes de Souza and S. Springer, *The Practice of Freedom: Anarchism, Geography and the Spirit of Revolt*, 65–84. London: Rowman and Littlefield, 2016.

Wilson, C., *Anarchist Essays.* London: Freedom Press, 2000.

Wise, M.J., 'The Scott Keltie Report 1885 and the teaching of geography in Great Britain'. *The Geographical Journal*, no. 152(1986): 367–382.

Withers, C.W.J., 'Towards a history of geography in the public sphere'. *History of Science* 37, no. 1(1998): 45–78. doi:10.1177/007327539903700102

Withers, C.W.J., 'Constructing the geographical archive'. *Area*, no. 34(2004): 303–311. doi:10.1111/1475-4762.00084

Withers, C.W.J., 'History and philosophy of geography 2004–2005: biographies, practices, sites'. *Progress in Human Geography*, no. 31(2007): 67–76. doi:10.1177/0309132507073537

Withers, C.W.J., *Geography and Science in Britain, 1831–1939: A Study of the British Association for the Advancement of Science.* Manchester: Manchester University Press, 2010.

Withers, C.W.J., D. Finnegan and R. Higgitt, 'Geography's other histories? Geography and science in the British Association for the Advancement of Science, 1831–1933'. *Transactions of the Institute of British Geographers*, no. 31(2006), 433–451. doi:10.1111/j.1475-5661.2006.00231.x

Zibechi, R., *Dispersing Power: Social Movements as Anti-State Forces.* Edinburgh: AK Press, 2010.

Index

Entries in *italics* denote figures.

Printed and bound by CPI Group (UK) Ltd, Croydon, CR0 4YY

24/10/2024

01778282-0012